BITTER
NEMESIS
The Intimate History of Strychnine

BITTER
NEMESIS
The Intimate History of Strychnine

JOHN BUCKINGHAM

CRC Press
Taylor & Francis Group
Boca Raton   London   New York

CRC Press is an imprint of the
Taylor & Francis Group, an informa business

CRC Press
Taylor & Francis Group
6000 Broken Sound Parkway NW, Suite 300
Boca Raton, FL 33487-2742

© 2008 by Taylor & Francis Group, LLC
CRC Press is an imprint of Taylor & Francis Group, an Informa business

No claim to original U.S. Government works
Printed in the United States of America on acid-free paper
10 9 8 7 6 5 4 3 2 1

International Standard Book Number-13: 978-1-4200-5315-9 (Softcover)

Library of Congress Cataloging-in-Publication Data

Buckingham, J.
  Bitter nemesis : the intimate history of strychnine / John Buckingham.
    p. cm.
  Includes bibliographical references and index.
  ISBN-13: 978-1-4200-5315-9 (alk. paper)
  ISBN-10: 1-4200-5315-9 (alk. paper)
  1. Strychnine--History. I. Title.

RM666.S8B83 2007
615.9'52--dc22   2006101010

Visit the Taylor & Francis Web site at
http://www.taylorandfrancis.com

and the CRC Press Web site at
http://www.crcpress.com

Contents

List of Illustrations   vii

Picture Credits   ix

Acknowledgments   xi

Note on Weights and Measures   xiii

Note on Nomenclature   xv

Introduction   xvii

1   Some Disadvantages of a Weak Constitution   1

2   Nuts   19

3   The Patient Generally Lies on His Back   33

4   M. Vauquelin's Lack of Fame   47

5   Perfidious Dutchmen Bark up the Wrong Tree   57

6   You Will Be Careful as to the Second Article   63

7   You Hold Him Down, I'll Pour It Down His Throat   83

8   Overture to the Sorcerer's Apprentice   97

9   The Fop, the Scotsman and the Opium-Eater   111

10 It Will Be the Test-Tube and the Retort That Will Hang Him   125

11 Shaken in Every Possible Way   141

12 Mrs. Dove's Brush with the Media   167

13 That Clever Dr. Letheby, So Ugly and Terrific   181

14 Tigers, Lions, etc.; Six Hundred Kilograms   201

15 The Blue Anchor Murder and Other Outrages   221

16 I Didn't Know It Was Used for Poisoning   239

17 Is There a Faceless Fiend?   247

18 Another Round of Pay Phone Hysteria   259

Bibliography   269

Index   277

List of Illustrations

Figure 1. Traditional medicine in India. Servant and Hakeem (physician).

Figure 2. *Strychnos nux-vomica*, after M.A. Burnett, 1847.

Figure 3. Strychnine; nux vomica seeds; Parke Davis & Co. strychnine sulphate for injection, circa 1900.

Figure 4. *Taking an Emetic*, by Cruickshank, 1800.

Figure 5. Eight leading French doctors of the early 1800s.

Figure 6. Knowledge of nux vomica reaches the West. Part of a manuscript in the Vienna state archives.

Figure 7. The ruins of the Chateau de Bitremont in Belgium.

Figure 8. Edmé-Samuel Castaing and his two victims, Auguste and Hippolyte Ballet.

Figure 9. Sir Robert Christison, Thomas De Quincey, and Thomas Wainewright.

Figure 10. A Victorian phrenology chart incorporating William Palmer as the criminal archetype.

Figure 11. A victim of tetanus.

Figure 12. The Old Town Hall, Rugeley, during the inquest on Walter Palmer.

Figure 13. Alfred Swaine Taylor and Dr. Rees.

Figure 14. Christiana Edmunds.

Figure 15. Thomas Cream.

Figure 16. 1904 Olympic marathon winner Thomas Hicks under the influence of strychnine and alcohol.

Figure 17. Who says there was no strychnine in beer? Cartoon from *Punch*, 1855.

Figure 18. Jean Pierre Vaquier and Mabel Jones.

Figure 19. Robert Robinson and R.B. Woodward.

Figure 20. The 1917 assassination plot. The three female conspirators.

Figure 21. Transvestite cabaret artiste Molly Strychnine with her accompanist Fuckoffsky.

Picture Credits

Figure 1: Reproduced with permission of the British Library, London, UK

Figure 2: Reproduced with permission of the Wellcome Medical Library, London, UK

Figure 3: Used with permission of Petra Laidlaw (also cover illustration)

Figure 4: Reproduced with permission of the Wellcome Medical Library, London, UK

Figure 5: Reproduced with permission of the Wellcome Medical Library, London, UK

Figure 6: Reproduced with permission of the Öster-reichische Nationalbibliothek, Vienna

Figure 7: Photograph by the author

Figure 8: Reproduced with permission of the Wellcome Medical Library, London, UK

Figure 9a: Reproduced with permission of the Wellcome Medical Library, London, UK

Figure 9b: Reproduced with permission of the National Portrait Gallery, London, UK

Figure 9c: Reproduced from Curling, Jonathan, *Janus Weathercock; the Life of Thomas Griffiths Wainewright*, 1938, location of original unknown

Figure 10: Reproduced with permission of the Wellcome Medical Library, London, UK

Figure 11: Reproduced with acknowledgment to www. publichealth.pitt.edu

Figure 12: Reproduced from the *Illustrated Times*, February 1856, from the author's collection

Figure 13: Reproduced from the *Illustrated Times*, 1856, from the author's collection

Figure 14: From the author's collection, original source unknown

Figure 15: Reproduced with acknowledgement to www. casebook.org

Figure 16: Reproduced with permission of the IOC/ Olympic Museum Collections, Lausanne, Switzerland

Figure 17: Reproduced from *Punch*, 1855, from the author's collection

Figure 18: Reproduced from R.H. Blundell and R.E. Seaton, *Trial of Jean Pierre Vaquier,* William Hodge and Company, 1929, original credits to Topical Press Agency and London News Agency, current copyright holder unknown

Figure 19: Reproduced with permission of Professor Jack D. Roberts, California Institute of Technology, California, USA

Figure 20: Reproduced from the *Illustrated London News*, 1917, from the author's collection

Figure 21: Reproduced with permission of Gert Volkmer and Tom Werkhoven, Amsterdam, Netherlands

Acknowledgments

A number of people provided valuable information or assistance at various points during the writing of this book. With apologies for any inadvertent omissions, I would like to acknowledge the following, some of whose contributions are further mentioned in footnotes.

Professor Peter Bladon, Luigi Bonomi, Trupti Desai, Professor Jim Fallon, Dominic Farr of the William Salt library in Stafford, Leslie Forbes, Stephen J. Greenberg of the National Library of Medicine, Bethesda, Pat Hall, Chris Hallam, Celia Hayley, Angelo Hornack, Petra Laidlaw, Runa Mackenzie, Lorraine Moran of Weatherbys, Professor Gerald Pattenden, Professor Michael Rogers, Philip Powell, Dr. Roger Snowden, Ian and Michaela Taylor, Professor Jos van der Maesen, Marc Vaulbert de Chantilly, Gert Volkmer; also the staff of various libraries including the Wellcome, the London Library, the Royal Society of Chemistry, The Royal Pharmaceutical Society, the British Library, The Linnean Society, the Guildhall Library, the Library of the Academie Nationale de Médecine, Paris, and several Edinburgh libraries.

Note on Weights and Measures

A *grain* was the smallest unit of weight in the imperial system. Originally based on the weight of a grain of wheat from the middle of an ear, in 1542 it was redefined in England as 1/7000 of a pound avoirdupois, which meant that there were 437½ grains to the ounce avoirdupois (the normal household imperial system still used in the United States and the United Kingdom).

Medicines, however, were usually dispensed using troy weights. In apothecaries' measures there were 20 grains in a scruple, 3 scruples in a drachm, 8 drachms in an ounce and 12 ounces in a pound.

The grain was used both in Britain and the United States, and in the early nineteenth century was also used on the continent concurrently with the metric system; however, the old French grain was not the same as the English one. The metric system replaced the grain for small weights throughout the world later in the century, with the legal world lagging behind the scientific.

One English grain was equal to 64.8 milligrams (mg). The old French grain was 53.11 mg, but appears to have been redefined after the metric system came in as 50 mg. (It is said that there were 25,000 different units of weight and measure current in France until the revolution, and these did not disappear overnight.)

Liquid measures were just as complex.

Note on Nomenclature

Strychnos is the botanical genus name for plants related to the nux vomica tree *Strychnos nux-vomica*. Nux vomica was powdered *S. nux-vomica* seeds containing strychnine, brucine and other substances.

Strychnia and brucia (sometimes bruschia) were the plant extracts containing strychnine and brucine, respectively, in a greater or lesser degree of purity and homogeneity. Strychnine and brucine are the pure active alkaloids. Most lawyers in the trials reported here used the terms "strychnine/strychnia" and "brucine/bruschia" interchangeably.

The strength of nux vomica preparations would vary considerably. Unprocessed nux vomica powder contained about 0.5–1% strychnine plus a similar amount of brucine. In 1890 the British Pharmacopoeia preparation of medicinal nux vomica contained half a grain of strychnine per ounce, or 7.4%, and was thus considerably stronger than the raw ground seeds.

Introduction

Fierce in dead silence, upon the blasted heath
fell Upas sits, the hydra-tree of death

Erasmus Darwin

One July morning early in the nineteenth century, Mr. Loudon, Mr. Strendenberg, and Mr. Darndels set off from the village of Batier in the inaccessible central highlands of Java.[1] Their destination was the valley called Guwo Upas, the home of the deadly Upas tree, so poisonous that it could kill by sight alone.[2]

Half a century earlier, in 1755, the explorer Foersch had found no trees growing within eighteen miles of the spot where the Upas tree was said to grow, and had described how local criminals were expelled from their villages to collect its leaves, and how very few of them returned. He witnessed the execution for adultery of thirteen village women by having the gum of the Upas tree injected into their breasts with a kind of spring-loaded lancet made out of bamboo.

The three explorers entered the secret valley, taking with them two dogs and a number of birds. After scrambling up a hill they were assailed by a nauseous smell, but on reaching the top of the ridge this disappeared. They gazed in astonishment at the scene before them; the oval valley, about half a mile in circumference, contained only the skeletons of human beings, tigers, peacocks, pigs and deer. Lighting their cigars, they attached one of the dogs to a bamboo pole eighteen feet long, and sent it down towards the valley floor. After fourteen seconds it collapsed on its back. A fowl thrown in expired before touching the ground. There was no smell of sulphur or of any recent volcanic activity, although there were two craters emitting smoke not far away.

In due course the explorers reported their findings back to Europe. Eventually it was realised that two different exaggerated legends, firstly of a deadly tree, and secondly of a deadly valley, had become conflated in the local mythology. The deadly valley was not a myth, but it gained its terror from carbon monoxide and other gases emitted from volcanic vents. Neither were the stories of the deadly Upas tree without foundation. It was not an emanation from the tree that killed people before they had a chance to touch it, but the leaves and every other part of this large tree, growing to a height of eighty or a hundred feet without branching, contain large quantities of a deadly poison.

Not long after the Javan expedition, and on the other side of the world, the poison that killed the village women was isolated in pure form from another plant closely related to the Upas tree. This substance, named strychnine, caused a scientific sensation. Its formidable properties seemed to promise the physicians a powerful new weapon to explore and control disease. Its uniquely powerful effects on the nervous system, closely resembling those produced by a powerful electric current, also promised them intimate insight into the functioning of the human body.

This book is the story of how these hopes were not fulfilled. Not only did strychnine fail to live up to its initial promise, not only did it prove a uniquely dangerous medicine, but it became the poison of choice for some of the most sinister of the subsequent century's murderers. How such an appalling poison managed to stay in the pharmacopoeias for a century and a half despite continuously mounting evidence about its lack of usefulness is just one of the questions that we shall try to answer. Today its reputation is for most people little more than as a shorthand for a highly poisonous substance; the one for example with which Richard Dreyfus armed himself when he went down in his cage to do battle with the giant shark in *Jaws*. But it has a far more intricate and chequered history than is encompassed by such a stereotypical image.

Strychnine, together with its close relative brucine, is found in a number of trees growing in tropical regions of the world, especially India and the Far East. The Upas tree, *Strychnos tieute*, is not the most famous or the most important of these. Its smaller and more widespread cousin, the nux vomica tree *Strychnos nux-vomica,* became the commercial source of the drug. Despite the colourful legends surrounding the formidably poisonous Upas,[3] it was the nux vomica or poison-nut tree that was truly to come to deserve the reputation of a dealer of death. This smaller tree cast the longer shadow, westwards.

Endnotes

1. The introduction is based on the account by Sigmond, 1837.
2. Legends of poisons that could kill by sight alone were common in some parts of the world. The Arab doctor Ibn Wahsiyyah, writing in the Tenth Century, said "There is a tree in India called sulūsūquir. When a lion sees it, then it dies without delay." An early reference to Strychnos? The name sulūsūquir is unidentified.
3. There were two kinds of Upas, the Upas Tieute and the Upas Antriar. Both are highly poisonous, but botanically and chemically they are unrelated. It was the strychnine-containing Upas Tiute which killed the village women, although the Upas Antriar is the larger tree. The poison of the Upas Antriar (*Antiaris toxicaria*) is not an alkaloid. It is related to digitalin, the heart poison from the foxglove.

CHAPTER 1

Some Disadvantages
of a Weak Constitution

*In the most obscure, difficult and dangerous maladies it is better
to hazard a doubtful remedy than to give nothing at all.*

E.R. Arnaud, 1650

Some years after the Java expedition, and in the wake of the
French Revolution,[1] there was founded in Paris by Ministerial
decree the Society of the Faculty of Medicine. A forerunner to the
French Academy of Medicine, it numbered among its membership twenty-six professors, sixteen associates and various other
savants. The Society set out with the goal of furthering the cause
of medical research and disseminating truthful medical knowledge throughout society, thus building a scientific edifice fit to
stand on the ruins of the old France alongside the new systems of
law, government and education now being set up.

Medicine was not the only scientific discipline that received a
powerful boost in France in these times; or, to put the case more
accurately, several important sciences did not yet exist in their modern form, and began to emerge as part of medicine. One of the new
emerging sciences was chemistry, effectively founded in its modern
form by the aristocratic Antoine Lavoisier, who after launching his
Chemical Revolution met his death at the guillotine. Lavoisier was
also one of the founders of another new science: physiology, the

1

systematic study of how the organism functions. Before his death, in collaboration with François Magendie, Lavoisier had begun to measure, for example, how much oxygen living animals require. These experiments were still at a preliminary stage when he was executed, but Magendie, who survived him, went on to become the leading physiologist of the pioneering days.[2]

Many physicians in other more theocratic countries considered it impertinent to probe the innermost workings of the body, and that medicine should only be approached holistically. The revolutionary spirit rejected this. The body was a machine, one whose workings could be probed with vigour but which were for the time being virtually inexplicable. The range of topics that people calling themselves physicians were now studying in France was extraordinary, for there were so many unanswered questions that almost any discovery in the natural world might have some medical application. By 1800 France had more scientists than any other nation had ever had. Some of the ideas tossed around were crack-brained, such as the belief in animal magnetism that had a powerful influence on science beginning in the late eighteenth century, generated a learned commission in 1826 to study its claims, then faded away to leave faint echoes even today in various kinds of quack products.

One of the members of this 1826 commission was Pierre-Éloi Fouquier, by this time an established physician of some repute at the Hôpital de la Charité in Paris. Very little is known about Fouquier's early life, except that he was born on July 26, 1776 in a hamlet on the Aisne called Maissemy, and that in his earlier years he styled himself Fouquier de Maissemy. This, his obituarist hastens to tell us, was not due to pretentiousness on his part, but to make it clear that he had no connection whatever with Fouquier-Tinville, the revolutionary prosecutor of dreaded memory, who had come from the same part of France. For Fouquier the physician was an idealist, determined to do good, not evil. Like nearly all French doctors of that era, he probably commenced his career ministering to the Republican army, but before long he was pursuing a pacific course of healing among the hospitals and libraries of the metropolis.[3]

Fouquier was not one of the founding members of the Society, since when it first convened he was only 24 years old, and he did not obtain his doctorate of medicine until two years later. But he soon became a member, and it is in its bulletins, beginning in 1811, that we first read of his experimental treatment of patients with a new and remarkable drug for paralysis.

A few years earlier, Fouquier's doctoral thesis had had the unexpected title *The Advantages of a Weak Constitution*. To explain why a doctor at such a time might write such a thesis, and how such a belief might in turn cause him to experiment on his patients by giving them a violently poisonous plant drug, we need to take a brief look at the state of medical knowledge at the turn of the nineteenth century.

The phenomenal complexity of the human body had long been clear to the anatomists, who since the Renaissance had been mapping its internal structures in great detail. But there was little or no understanding of what those extraordinary organs actually did. As a result, well into the nineteenth century, medicine continued to be dominated by simplistic ideas often held in opposition by competing sets of doctors. For example, Galen, the Roman physician whose ideas dominated medicine through the Middle Ages and well beyond, had believed that nature produced a plant to cure every malady. A closely related and widespread idea is that every plant has been put on the earth with a specific use in mind. These ideas were running strongly in 1800, and persist today.

It is true that many plants do have extremely useful medicinal properties, and even now, about one third of all newly introduced drugs are plant products, or closely based on a compound first isolated from a plant. An outstanding example which had been around since the middle of the sixteenth century was quinine. This is named after the Countess of Chinchon, Dona Francisca de Rivera, who became ill in South America with malaria in the 1630s. In desperation, her doctor turned to the local Indians for help, and was told to give her the bark of a tree, now called Cinchona. Quinine is the active principle of the tree and is a genuinely useful drug. But such a dramatic success had several drawbacks. It helped to reinforce the complementary views, firstly that every plant was useful, and secondly that every disease had a natural cure; and, because there was no scientific basis for distinguishing between the many infectious diseases, of which malaria is only one of the more important, the belief persisted that quinine could treat any fever. Furthermore, in the days when scientific knowledge was so rudimentary, the only property that seemed characteristic of quinine and could seem to account for its medicinal properties was its bitterness, and by extension, medicinal powers were assigned to any bitter plant extract. The stage was set for the appearance on the scene of the extraordinarily bitter nux vomica.

Galen also believed that virtually all disease is the result of disorders of the blood, the principal organ of the body. This belief was a natural consequence of his doctrine of the four humours and led on to the main medical treatments of the eighteenth century which involved various ways of drawing off excess or diseased blood by leeching or cupping (sucking blood out with the vacuum created by a heated wineglass). Taking into account the other main contemporary stratagems; purging, blistering, vomiting and the lancing of boils, it is clear that virtually all the techniques meted out by the eighteenth-century physician had to do with expelling noxious matter from the body. The belief that blood is the organ in which all important metabolic changes take place was widely held until the middle of the nineteenth century and influenced the views of forensic scientists among others. It was not until more understanding of the immense importance of the mysterious liver and other organs began to dawn as the century progressed, that these ideas could be modified.

By the opening years of the nineteenth century, French physicians were probing the organism in new ways. New questions were beginning to be asked and new techniques tried, although the continued lack of real understanding of chemistry and physiology meant that for the time being, a new generation of simplistic explanations held sway. One important preoccupation at this time became weakness, or debility. Such a sweeping generalization was unlikely to advance medicine very much, for a patient can suffer loss of strength through a variety of causes; injury to the brain or nerves, or a hereditary metabolic disorder, or a multitude of other causes. And one must never forget that until Pasteur's work of the 1860s, doctors remained in complete ignorance of the cause of the many weakening infectious diseases; although it would be wrong to assume that they could not therefore distinguish these one from another by studying their symptoms. Nevertheless, one of the most important disciplines that was emerging was neurology, with its understanding that the impulses that drove the muscles seemed to emanate from the brain and spinal column and move electrically through the motor nerves, while sensations of pain and pleasure that originated in the different parts of the body passed in the opposite direction along a different set of nerves entirely.

Since we know so little about Fouquier's early life, it would be idle to speculate what might have led him to choose as his 1802 thesis the question of why weak-bodied people might be superior to those apparently more robust. But let's do it anyway. The engraving of him as a mature man (Figure 5) shows someone apparently sound in wind and limb. We do not know anything about his parents, or whether he had siblings, but it is impossible to avoid the feeling that he must have been dominated as a child, or perhaps that he saw his mother being bullied.

Physically weak people, he says, although despised by the strong, have greater spirit since it is well known that intellectual and moral strength is inversely related to physical hardiness. A weak constitution is the result of the predominance of the nervous and lymphatic systems, and is characterized, particularly among women and children, by a greater volume of the brain and nerves, by the rapid effect of medicaments, greater sensitivity to pleasure and pain and a more active imagination. Infants grow out of it, but women are condemned to permanent inaction and subjection. Some men, those with an artistic temperament, had such a constitution, and tended to avoid the diseases to which those of a robust constitution such as labourers are prone, especially the diseases of excess. When strong men fell ill with an obscure malady, they were prone to aggravate it by their vanity and aggressive attitude, thus ignoring it until a doctor was no longer able to help. It was well known that if diseases of confinement and other female illnesses were discounted, women were less likely to succumb to fevers than men. When a weak person of higher moral calibre fell ill, the phlegmacies were more likely to be of a mucuous nature and their fevers slow and intermittent, and if a doctor made a mistake in treating them, he had time to correct the error before it got out of hand. Intemperance, especially debauchery, was the greatest cause of illness, and the lifestyle of Northern Europeans was killing them because they worked too hard physically and ate too much meat. The French, English and Germans were the unhealthiest people in the world, and the moment of greatest attainment of an empire was always the moment when physical degeneration begins among its population.

We may laugh, but as ever, there are grains of truth. Big, muscular people are indeed more prone to some maladies. Nineteenth century industrial societies were indeed extremely unhealthy. The besetting sin of nineteenth-century medicine was not in the things it got diametrically wrong, but in the way that it applied rash generalizations to illness and other nebulous concepts, and larded each generalization with irrational fears about the working classes, sex, national degeneration and other dreads that were to dominate the century.

Having set out his stall with the argument that weaker people of a more nervous disposition enjoy various advantages, in the appendix to his thesis Fouquier begins to contradict himself. Vigour, pressed to the extreme, he says, merits all the strictures I have made on it. But if the nervous and lymphatic systems are too dominant, the power of the organs is weak and disorganized. Just as overuse of the muscles can paralyse the brain, depriving it of all moral action, so can the work of the spirit destroy physical force. A shock or sadness can deprive the limbs of action. Those who are enervated by excessive study, like the victims of masturbation, are subject to collapse and can die of nervous phthisis.

To languish as a valetudinarian is not to live, and Fouquier is resolved to research the best means to prevent this deterioration and fortify the constitution. How to bring this about? The clue lies in his correlation of meat-eating with disease. It is in his plan, he says, to show that vegetable nourishment is the basis of a robust constitution. The country peasants of France thrive on a diet of plants, whilst a crowd of meat-eaters in the cities fall victim to religion and idleness. (Come, come, Fouquier. Does meat cause too much vigour or doesn't it? Make your mind up.) What fruit can rescue them? Among the plants, he says, there is a nutritive power not found in meat, and it is necessary to distinguish among the different alimentary substances of their juices.

Fouquier was a true child of the revolution. He rejected spirituality and claimed to adhere to the doctrine of simple good sense. Science came first, and he was prepared to risk his patients to forward the course of medicine. "Extreme timidity will never permit a fortunate outcome," he said, and began to cast around for a new tool that would restore the fitness of patients whose constitutions had become dominated by their nervous and lymphatic systems, thus depriving them of their muscular strength. Such a product would have to be ferocious in its effect, and would have to come from a plant. Before the end of the first decade of the new century, he thought he had found it.

Although known since the Middle Ages and haphazardly toyed with by doctors, particularly in Germany, for the previous two centuries, nux vomica, the powdered nuts of a tree which grew in India, was considered so unreliable and dangerous that until now no French physician had seriously considered it as a viable medicine; its reputation was almost entirely as a useful poison for unwanted dogs. Some German physicians, using it mostly against dysentery, had accused the French of timidity in rejecting it. But now, the more that Fouquier studied what reports there were of its effects, the more he considered that in the right hands, and with a sufficiently careful scientific approach, its possibilities were almost limitless.

On May 24, 1811, Fouquier delivered a lecture to the Society describing the results of his first experiments on patients using the new drug, but no record survives of what he said. We have to wait another five years before a lengthy report, split into three parts, was published in the Bulletin of the Society for 1816–1817. The report describes the results of administering an alcoholic nux vomica extract to sixteen patients suffering from paralysis. If Fouquier had hoped to further explore his theory that members of the intellectual classes could be reenergized using his new plant medicament he was disappointed, for the clientele of the Hôpital de la Charité was uniformly proletarian; a carpenter, a weaver, a dressmaker, a cobbler and so on. There are faint echoes of his earlier theories in the way he describes some of his patients as having a lymphatic constitution, but in general the ideas expressed in his thesis are lost sight of, and before long he is trying the effect of his nux vomica extract *tout court.*

Accordingly, on November 1, 1812, Fouquier administered the alcoholic extract of four grains of it to a M. Burion, a 34-year-old upholsterer normally resident at no. 88 Rue du faubourg Saint-Denis. Burion, Fouquier informs us, was of a lymphatic temperament and had suffered from scrofulous swellings[4] of the joints since the age of seventeen. For several months his pedal extremities had been paralysed, and the paralysis had now spread as far as the loins, affecting first one side of the body, then the other. By September 1812 the paralysis had reached his abdomen and he was no longer able to stand upright. Constipated and incontinent of urine, he was obliged to keep to his bed. A fortnight's treatment with aromatic baths and various other medicaments had been unsuccessful. There was total loss of feeling in the limbs, which treatment with blistering plasters had not allayed.

For the first three days the nux vomica, given in doses increasing to eight, then ten, grains per day, had no effect. Then on the fourth day, when a dose of twelve grains was administered, Burion felt during the night a kind of shock which seemed to originate in the right flank and spill over into the rest of his body. Two hours afterwards, having suffered several smaller shocks, he experienced cramping in the stomach. These spasms reappeared several times a day, especially when he tried to move in any way. They were often accompanied by convulsions of the abdomen and even of the jaw, usually followed by a kind of numbness.

By November 8th, the legs were capable of some exercise, and later in the month his control of urine, which became clear again, returned. He could walk, first with crutches, then with a stick, although the force of his abdominal muscles had not completely returned. On December 5th, having consumed a total of 314 grains of nux vomica and abandoning the stick, he left the hospital in a state of perfect cure.

This is the first of Fouquier's reported cases, and the one that gives the clearest description of a claimed success. Some of his other cases are far less clear-cut. One fact that sticks out like a sore thumb even to a nonmedical reader is the obvious dissimilarity of the causes of the paralyses that he treats.

M. Rousseau, age 60, clearly seems to have suffered a stroke, while M. Vanhove, a 62-year-old cobbler, is suffering paralysis after a fall, and Charles-Martin Paris has a lower-back injury, probably a slipped disc. On the other hand, some cases seem to have been caused by disease, such as Burion, with his scrofulous swellings, and François Janhin, a weaver, who had suffered for many years with rheumatic pains in the right arm although he averred that he had never had a venereal disease. Madame Ponsin, age 31, is there because on August 20, 1815, she accidentally drank some liquid intended to kill flies. Since this did not kill her during the period of more than two months that it took for her to be admitted to la Charité, it is hardly surprising that after lying in a hospital bed for a further two-month period she felt well enough to go home, despite every effort of Fouquier and his colleagues to finish her off with their extract of nux vomica. In several other cases the patient already seemed to be getting better when the cure was started. Marie Fredoy, a dressmaker, had suffered for fifteen years from paralysis of the lower limbs caused by what she described as an aggravating life. When the paralysis accompanied by headaches recurred, she entered the hospital, where she was treated with a blistering plaster to the back of the neck and twenty leeches. After several days, she was beginning to recover movement of her abdomen and thorax, and it was at this point that Fouquier began his treatment. Nux vomica was given for a full three months, accompanied by the usual *bouleversements* and *tressaillements*, after which period she could walk with a stick but at this point "I lost her to view, and my treatment ceased."

The reports by Fouquier and his junior colleague Bricheteau concerning their cases are hardly models of a scientific method and convey a graphic sense of what life was like in a hospital ward of the Napoleonic period. There are very few claimed cures that are not complicated by the patient disappearing before the doctors have a chance to complete their treatment, or failing to be conclusive for some other reason. M. Vanhove, for example, at first shows a remarkable improvement, but then has the misfortune to be given nux vomica pills intended for another patient at the same time as he is receiving enemas of nux vomica on his own account. This double-ended administration had the unfortunate effect of causing violent convulsions, after which there was no further improvement, and at the time of writing he remained in the hospital ward still subject to partial paralysis. Bricheteau states that, in his opinion, the man has cerebral congestion as a result of his fall and that if only he had received the drug soon after this accident, he would by now be cured.

There is also the knotty problem of trying to decide which of the patients may or may not have been malingerers. Consider the dates of Fouquier's experiments. The year 1812 is that of Napoleon's retreat from Moscow, and we are in the midst of war. Most of his patients are male. One or two are described as former soldiers, and several of those that are not are carters, wheelwrights and so on; occupations that would have seen them dragooned into service with the *Grand Armée*. Some of them sound worn out and were doubtless looking towards a spell in a comfortable hospital as a palatable alternative to the excursion to Moscow and back. Until, that is, they experienced the full impact of the treatment that Fouquier had lined up for them, when they tended before long to beat a hasty retreat, crutches waving in the breeze as by a supreme effort of will they powered themselves back onto the street.

> Case 4. Guillaume Pigny, age 50, porter, strongly built, has always enjoyed good health except for two bouts of pneumonia and one of pleurisy. At the end of February 1814, Pigny was soaked by cold rain during a forced march. I could not decide to what extent this caused the illness that affected him soon afterwards, but during this time he complained frequently of a general lassitude which he attributed to an excess of work.

Describing the case of Maurice Cillière, a 64-year-old carpenter, Fouquier reports:

> I learnt that this man had neglected to take the remedy properly, less through insouciance than because a cure would disqualify him for admission to a hospice. I had him watched. The usual dose of nux vomica, which apparently he had never before taken, then produced a general paralysis lasting 8 hrs, and which frightened the assistants as much as the patient.

M. Paris, a wheelwright of no. 18 rue du faubourg du Roule, also showed improvement in his condition as the level of dosage was gradually increased, but again, "The improvement was especially marked after an extraordinary dose of medicament was given inadvertently." Not long afterwards, Fouquier suspected dissimulation on the part of the patient, and therefore "I resolved to subjugate him to a regular treatment, and the measures that I took allowed no possibility of evasion," as a result of which Paris said he wanted to be off, and left the hospital walking without assistance on July 28, 1813 able to resume work.

Numerous of the claimed cures were thus equivocal. The good sense, or cowardice, of the patient was often a contributory factor.

Case 12. Joseph Merré, a locksmith. Of strong constitution, born of healthy parents and of lymphatic/sanguine temperament. Since childhood he had been subject to headaches and in 1788 suffered a bilious-putrid fever, which caused the right members to become enfeebled. He recovered after three to four months but the right side remained rather weak. In July 1810 he suffered flashes in the eyes, buzzing in the ears and dizzy spells; the right side was considerably weaker and he had difficulty speaking. He entered the hospital with a firm stride but by the following day paralysis of the right side was complete. After three months of conventional treatment, there was some return of movement. In September I applied a vesicant to the nape of the neck, and the patient ascribed his subsequent improvement to this. By January 14th he could get out of bed but not dress himself; the arm was stiff and immobile. I began nux vomica powder at the rate of ten grains per day, increasing the dose to fifteen and then to twenty. For eight days there were no effects worth reporting, but then began violent spasms in the affected parts and although I reduced the dose to half, they persisted several times a day, especially at night, with strong uncomfortable cramps on the right side. Toward the end of the month, I substituted alcoholic extract of nux vomica; the spasms continued. The movements of the arm and hand became stronger and surer. On February 10th he could walk the length of the room, get up and dress unaided. He began to suffer from coughing fits which in the night provoked a spasm of the diaphragm so violent that he thought he was going to suffocate. I suspended treatment in order to recommence it later, but the patient was convinced that nux vomica was the cause of his cough, refused to take any more, and the cure remained incomplete.

Rousseau, too, remained incompletely cured after he suffered continual insomnia, culminating in a kind of night-time delirium which rendered him insane, after which he refused any more treatment.

Fouquier was treating his patients with nux vomica powder, the ground-up seeds of the nux vomica tree, or else with an alcoholic extract of the powder; given usually by mouth, or in the case of the unfortunate M. Vanhove, by enema. The extraordinary symptoms suffered by his patients were due to the presence in the powder of the violently poisonous alkaloid strychnine, together with its close relative the somewhat less potent brucine.

The neural effects of strychnine were first described properly by DeBarenne in 1915. Once absorbed and circulating in the blood, the strychnine molecules lock onto receptors in the nerves controlling the muscles, preventing the functioning of the chemical messenger (glycine) that normally controls them. The nerves become hyperexcitable and as a result the muscles go into immensely powerful and excruciatingly painful spasms. When a muscle contracts, the opposing muscle should relax under the control of a reflex passing through the nerves leading from the spinal cord. But strychnine blocks the nerve transmission to both sets of muscles, causing them both to contract violently. The muscles affected include those needed for respiration and so the victim is prevented from breathing. Death is caused by suffocation; the muscles continue to contract until exhaustion causes a temporary respite, the victim dies, or the strychnine wears off and he recovers. There is arching of the back (opisthotonos), and the characteristic *Risus sardonicus*, caused by contraction of the facial muscles. The eyes are partly closed, the forehead furrowed, the nostrils flared and the lips pursed, thinned and stretched; the Latin label is somewhat misleading, for the expression is really one of pain, anguish and fear. Strychnine does not cross the blood–brain barrier, so conscious thought and the mental facilities are not affected. (An additional factor is that glycine receptors are especially distributed in the brainstem and spinal cord, whereas other types of receptors predominate in the cerebellum and forebrain; hence strychnine's lack of effect on the conscious functions.) Some South American plants closely related to the nux vomica tree produce curare, which is closely related chemically to strychnine but has a quite different physiological effect, leaving the glycine receptors unaffected but instead blocking acetylcholine receptors and thus paralyzing the skeletal muscles.

Although it does not affect consciousness, one symptom universally reported by those who have taken strychnine and survived, or have taken it and are dying, is a feeling of impending death. Psychological feedback from the way that the face is forced by the drug into an expression of fear and dread might presumably have something to do with this, but the feeling of having the life squeezed out of one's body by a giant invisible vice is surely enough in itself.

There is only one natural disease that resembles the effects of strychnine poisoning, and that is tetanus, a disease that results from contamination of a wound by the bacterium *Clostridium tetani*, the spores of which are widespread in the soil. In a closed wound, the bacteria produce the toxin tetanospasmin, a powerful nerve poison resembling strychnine in its effects.[5] The toxin is absorbed by the nerve fibres in the vicinity of the wound, and circulates in the bloodstream whence it can be absorbed by nerves in other parts of the body. It then migrates up the nerve fibres. Like strychnine, the toxin does not cross the blood–brain barrier, and so does not affect consciousness. The jaw, face and head are affected first because the nerves controlling their muscles are shorter and the toxin has a smaller distance to migrate. Therefore the first symptom, typically appearing about a week after first infection, is usually lockjaw, or trismus. Treatment of severe cases is not very effective and the mortality rate even today is about 50%. These symptoms are superficially very similar to those of strychnine, but the onset is much more gradual. The medical distinction between strychnine poisoning and tetanus was extensively gone into forty years after Fouquier's experiments, during the Palmer trial, as we shall see.

In 1895, a graphic account of what it feels like to be poisoned by strychnine was given by W.T. Harris, a doctor practicing in South Africa.[6]

In January, 1893, it happened that I had for a few weeks been in the habit of taking an occasional dose of one of our stock dispensary mixtures—a tonic containing, amongst other things, a fair dose of strychnine...on the morning of Tuesday, January 10th, I went into the dispensary before the dispenser had arrived, to take a dose of the tonic I have alluded to. It is kept in a concentrated form, the whole bottle containing five drachms of the liq. Strychninæ B.P., and each ounce of the diluted mixture five minims. Somewhat carelessly I poured out sufficient to make an ounce and a half, and filling up the measure-glass with water, drank it off.

I at once noticed a much more intensely bitter taste than was usual; for although this characteristic of the drug may be detected in very dilute solutions, it seemed increased tenfold, as indeed it was almost, as I shall presently show. I immediately asked the porter if he knew when the mixture had been made up, and he replied that it had been done on the previous day, but as yet none had been dispensed from the bottle.

I did not know quite what to do, and my first impulse was to take an emetic; but, as the swallowing of saliva lessened the bitter taste every minute that I hesitated, I persuaded myself that the difference might be only fancy. I had made a good breakfast, and was loath to sacrifice my mutton-chop and upset my stomach, only to be laughed at; for how could a large stock-bottle be made up so improperly that an ordinary dose would do me harm? And was it not ready to be dispensed for a number of other people? I shook off my fancies therefore, and going into the consulting-room, rang the bell for out-patients and went on with the morning's work.

Fifteen minutes elapsed, and I began to feel very restless. An indescribable nervous sensation came over me, as if there were rope pulleys running down to my extremities, which were gradually being drawn tight. I had to make an effort to prevent my mouth closing too soon as I spoke, and to dig my pen into the paper and write thick, as if to form a fulcrum over which to lever my hand along the pages, while a contra-force in my arm strove to dash the pen to the floor.

Fortunately there were but few patients to see that morning, and I had just finished them, when at a little before eleven o'clock, Dr. Considine came in. I at once told him that I felt very strange, and feared I had taken an overdose of this strychnine mixture. He laughed and said I was nervous, knowing that we both had often taken the same medicine with impunity. He then commenced talking on some topic in which we were usually both interested, when I broke in abruptly, saying, "I feel I cannot sit still and talk; let us go round the wards."

We started through the principal male ward, which is a daily routine, and one always of interest, and generally of pleasure. But the simple round on that particular day seemed then in fact, what it still appears to memory, a dreadful nightmare. My limbs were throwing off the control of will, and moved erratically; when I wished to go on, my legs stopped, and when by a violent effort I forced them to proceed, I could not pull up to a standstill without walking against a bed to steady myself. What I said or did I cannot remember, but I managed to get along somehow, though feeling as if head, hands and legs belonged not to me but to three separated individuals, like a mechanical doll that has had all its limbs pulled with a jerk of the string. At length we returned to the top of the ward, when, feeling a paroxysm down my back, I said to Dr. Considine, "I am really very ill. I feel sure I am suffering from strychnine poisoning." I had taken six-tenths of a grain.

I remembered that half a grain had caused death. I must prepare to die—to die fearfully, to die soon...the simple fact to a man that *he* is to die is a heavy blow for the strongest will or the stoutest heart...the thought was horrible! In the fullness of strength to be delivered over to the power of this chemical monster, this nitrogen Nero, this molecule Moloch, this pent-up force which, like the genii of Eastern fiction, liberated by the fisherman Science from the vase of centuries, wreaks its frightful energy; latent under the guise of a harmless-looking crystal, but more death-dealing than dagger or dynamite...the deadly drug seems to revenge its former subjugators when once it gets the upper hand...

"Shall I have an emetic?" I asked, and Dr. Considine said, "No, it is too late; take sixty grains of chloral. Now go to the ophthalmic room, and smoke hard, if you can manage to." I turned to go. "You'll come soon; do not leave me for long," I said. "I'll come immediately, and not leave you, however long it is, till you are better." The words imbued me with new courage, though, as he told me afterwards, he feared the worst, and only stayed to get chloroform, morphia and a hypodermic syringe, should they be wanted.

I got down the passage, lay on the couch, and tried to smoke, but there was no rest possible; it was like lying on the felt floor of a Turkish bath. As one flinches there from the heat of contact with surrounding objects, so here every touch sent a tetanic convulsion through me. I could not rest; should I get worse and have opisthotonos? Would the chloral stop, or only stay, the action of the poison? Was it to be a reprieve, or only a respite? I started to my feet and got over to the book-case, to see what Taylor's "Jurisprudence" said as to the oncoming of symptoms and the period of danger. The book was not there, and I remembered that I had taken it over to my residence. How was I to get across the garden to my study, sixty or seventy yards away? What should I do if I met anyone, how afford an explanation with every muscle on the work, and feeling unable to articulate? I could only hope to get in and out unobserved, for I felt that not only would any attempt to explain bring on a paroxysm, but that I should cause the greatest alarm by my appearance. I started, and how I steered myself across is a problem still. I ran in jerks and jumps, just as a drunken man makes a dash from one lamp-post to another.

I regained the room in the hospital, and, steadying myself between couch and table, turned to the accounts of strychnine poisoning, feeling "a trembling of the whole frame" and "impending suffocation" as I hurriedly glanced at those very words, so well describing my own symptoms. The accounts were conflicting...at last my eye fixed on this; "In fatal cases death generally takes place within two hours." To that I pinned my faith, for it was nearly twelve o'clock, and every moment was a step towards safety. I was now able to lie down, the chloral was taking effect, not as a hypnotic, for I was never more wide awake, but I could feel it "Coursing through the narrow straits and alleys of the body" with a gentle glow, and the spasms were abating.

Dr. Considine, who had been coming in and out, now settled down beside me, his cheery words being an important factor in tiding me over the next hour. He did everything to divert attention from myself, relating how he had once taken an overdose of strychnine, and had used tobacco as the only antidote available. The question of more chloral was considered, but no more was given, as I had had a large dose, which he rightly thought would prove sufficient; and more was at hand at any moment had acute symptoms returned. As is usual in these cases when once poison is eliminated, I felt but little subsequent effect beyond some weariness after the shock.

I have now come to the end of my narrative, in which I have endeavored to portray faithfully the sensations, mental and bodily, that I went through. That I did not get worse, and actually recovered after only the one large dose of chloral, I attribute to my generally good constitution, to the fact that I had habituated my system somewhat to the action of strychnine by having taken for a week previously medicinal doses of it two or three times a day, and to the prompt administration of the antidote at the critical moment, when the symptoms were coming to a climax. I do not know of any other case on record where a medical man has been able to describe the effects on himself of a large dose of this poison, and Dr. Warner, who took less than I,[7] never lived to tell the tale.

Dr. Harris was wrong to attribute some of the credit for his survival to his habit of taking strychnine on a regular basis. It is not an habituating poison, like arsenic, nor is it a cumulative poison like lead. His account shows how by the end of the nineteenth century, strychnine had developed such an undeserved reputation as a useful medicament that two qualified doctors could consider it healthy to raid the dispensary and take it on a daily basis.

One of the questions that this book will try to answer is how such a deadly poison gained such an undeserved reputation. A contributory factor was the scientific climate in early nineteenth-century France. Despite the equivocal nature of most of his results, by the time that Fouquier published the full account of his trials, other French doctors, anxious not to miss out on the excitement generated by the new drug, had already begun to experiment on their own patients. Some of them were also quick to try to steal the credit of Fouquier's discovery. Many jumped on the band-wagon, and by 1820 it had begun to roll, rather fast.

Endnotes

1. It was founded on the 12th of Fructidor in the year VIII, that is to say 28 August 1800.
2. Olstead, 1944.
3. Requin, 1852; Busquet, 1927–36, vol. 2.
4. Scrofula is a form of tuberculosis, but "Scrofulous swellings" might not necessarily have been caused by scrofula.
5. Udwadia, 1996.
6. South African Medical Journal, April 1895:341.
7. W.C. Warner was a Vermont physician who accidentally took between a quarter and half a grain of strychnine at the age of 39 and died within 20 minutes. The apothecary gave him a bottle of strychnine in mistake for morphine, and although it was clearly labelled, he tore off the top of the wrapper and drank some of it down. Note the small fatal dose, the smallest recorded for an adult, although this information may not be reliable (*Medical Examiner*, 1847, vol. XXVI, p. 309).

CHAPTER **2**

Nuts

The orientals do not confine themselves, as did Mithridates, to make a cuirass of their poisons, but they also made them a dagger....With opium, belladonna, bruceaea, snake-wood and the cherry-laurel, they put to sleep all who stand in their way.

Alexandre Dumas, *The Count of Monte Cristo*, 1845

This is the story of how nux vomica reached the west. The centuries following the fall of Rome have often been dismissed as the "Dark Ages," but across the globe trade between peoples continued. Even when Venice and other Mediterranean cities rose to prominence at the end of the Middle Ages, sea trade was by no means confined to ships going to and from western Europe. Marco Polo (not always reliable) claimed that for every load of pepper travelling to Christendom, a hundred such, "Aye, and more too," travelled to the port of Zaytoun (Tsüan-chow) in China. During these long centuries, the Arabs controlled the main trade routes by which drugs, spices and all other kinds of goods reached the West, either overland through Egypt or through the Red Sea and so on to Salerno in the south of the Italian peninsula and thus to Venice, where in due course "Marvelous palaces rose beside green canals"[1] as a result of the trade in exotic drugs.

The Arab philosophers took over custody of the classical texts concerning medicine and other scientific knowledge. Greek and Roman writings on herbal medicine by Galen and others were copied into vast books which also incorporated the Arab physicians' own discoveries about the natural world. Two notable and massive compilations were the *Royal Book of All Medicine* (al-Kitab al-Malaki) by Ali ibn Abbas al-Majusi, known in the West as Haly Abbas, who died in 994, and slightly later the *Canon of Medicine* by ibn Sina known as Avicenna, who died in 1037.

Haly Abbas was born into a Zoroastrian family from the Persian city of Ahwaz. He practised medicine in Baghdad where he served as physician to Caliph Adud al-Dawlah. The Caliph is best remembered today for having built in 979 AD the shrine to Husayn, the most important martyr of the Shi'ite muslims, at Karbala in Iraq, but he also founded a hospital, the Adudu hospital, where Haly Abbas healed the sick and wrote his book. In Europe, his compendium became known as the *Liber regius* or the *Pantegni*.

It is deep in the pages of this vast medical compendium that there is the first recorded mention of some mysterious nuts reputed to have powerful medicinal properties. The nuts or seeds were about 2 cm. in diameter, very hard, greyish-white and satiny with a covering of hairs. By the time they had completed their months of travel from their place of origin, their interior was dark-coloured and horny. Known in Arabic as *Aiche rhorab* or *Khubz al ghurab* (crows' bread), they appear to have been unknown in the classical world. They later became known in the West as *nux vomica*.

By 1100 AD, Salerno in particular had become a centre of knowledge of the Arab science. The Benedictine monk Constantine the African instituted a programme of translation of the Arab works. For this he mostly employed Jews such as Isaac of Salerno, who would translate them into Hebrew first and then into Latin.[2] This led on to the writing of smaller condensed works, which were easier for monks to copy, carry about and consult. By the thirteenth century knowledge of the vomiting nuts, or *Nucis vomicae*, reached England in the *Antidotarum* of an author calling himself Nicolaus of Salerno or Nicolaus Praepositus,[3] writing between 1200 and 1250. Links between Salerno and England were particularly strong around 1200 AD; English monks, notably Adelard of Bath, travelled to Southern Italy.[4]

The origin of the nuts was certainly known to the Arabs, for at some point the trees from which they came were imported to the Middle East and cultivated. But further west in Europe, their source remained a mystery. It was not even certain that they were nuts at all. The early Western explorers, following in the footprints of Marco Polo, brought back little in the way of accurate descriptions of plants. They were adventurers whose concerns were to stay alive and to enrich themselves.

A notable exception was the Portuguese Garcia da Orta who, in the middle of the sixteenth century, gave the first description of a highly poisonous wood called *Lignum Colubrinum*, or snakewood, which was later shown to be the wood of either the selfsame tree from which the nuts came, or else of one of its close botanical relatives.

Da Orta was the son of a Spaniard, Fernando da Orta, and his wife Leonor Gómez, who were among five thousand Jews expelled from Spain as the result of a royal decree signed by King Ferdinand and Queen Isabella on March 31, 1492. Their son Garcia was born some time shortly after 1500 and studied to become a physician. But by the 1530s he realized that the power of the Inquisition was beginning to spread to Portugal. At this time, Vasco de Gama and others had opened up the sea route around the Cape of Good Hope, and were beginning to break the stranglehold of the Arabs over trade between the Indies and Europe. On March 12, 1534 da Orta took ship aboard the Rainha to the new colony of Goa, where he began to study the plants of the Indian subcontinent and the drugs derived from them. The results of his extensive studies were published in 1573 as his *Aromatum Historia,* one of the first three volumes to be printed in the East Indies.[5]

Concerning Snakewood or Lignum Colubrinum

Not only is this wood, or rather its root, effective against the poison of animals which bite and strike, but also its dust is believed to kill stomach worms and to remove pimples, skin eruptions and scabs, and to cure what they call the choleric disease (the natives call it *Mordexi*). Likewise they say it is a useful remedy against the passage of fevers when an ounce of it is administered, only, however, after it has been pounded and soaked in water and after much bile has been purged through vomiting.

That this root is indeed effective against snakebites has been discovered in this way. On the island of Ceylon there is a kind of snake that is furnished with a crown (the Portuguese call them *Cobras* after a small goat, but we call them *King*), and that is extremely poisonous. In like manner there is a kind of animal which is the size of a ferret and which is the mortal enemy of this snake; they call it *Quil* or *Quirpile*. Whenever this animal is about to bite the snake, it bites that root (which grows there in profusion) in that part where it has been exposed; indeed some part of it juts out from the ground. When it has bitten the root and sprinkled its front feet with saliva, it first strokes its head and then the rest of its body, then it attacks the snake and does not let go until it has killed it. But if it cannot defeat the snake in that battle, it has recourse again to the root, against which it rubs itself, and then returns to the battle, and thus it bites the snake and kills it. Enlightened by this spectacle, the Chingalae (the inhabitants of Ceylon) have discovered that the root resists poisons.

This root is pounded and given to those who have been struck by a snake; it is also rubbed onto the skin in a hygienic fashion and is sprinkled on wounds. They say that this root grows in many other places and on the island of Goa.[6]

Da Orta's snakewood, or something closely resembling it, had been used from time immemorial by the Malays as a medicine, and they may have exported it to China in 776 BC. It was so highly valued in India during the eighteenth century that it rarely reached Europe.[7] When it did, it received the name in French of *bois du couleuvre*.[8] The aggressive habits of the mongoose thus seem to have played a part in building up the reputation for medicinal usefulness of these mysterious woods, as well as the nuts that came from the same or related trees. Knowledge of how poisons worked, and how to protect oneself against them, was still inextricably mixed up with magic and superstition. A footnote, presumably inserted by Clusius, the translator of da Orta's book into French, says that in 1564 in Salamanca he was shown a fragment of snakewood, the length of three fingers, sent by a very distinguished man as a gift to his father (together with a splendid bezar stone and some small containers made of tortoiseshell). These things, he asserts, provide wondrous protection against poisons. It is ironical that such deadly poisonous plant materials built their reputation not only on their reputed medicinal properties, but also on their claimed power as antidotes to other poisons.

By the time da Orta wrote his book, the church authorities were doing all they could to prevent the dissemination of the new scientific knowledge. When he died, probably in 1568, it was just in time to escape the arrival of the news that his sister Catarina had been burnt at the stake in Lisbon. da Orta's timely death must have outraged the authorities, for they ordered that his corpse be disinterred, burnt, and the ashes thrown into the river Mandovy. The same incendiary fate awaited most copies of his book, which are now very rare.

As knowledge of the new Eastern drugs trickled westwards, mentions of nux vomica and lignum colubrinum began to appear in the herbals, or books describing different plants and herbal medicines, that were written, first on the Continent, then in England. The first Continental pharmacopoeia was published in Florence in 1498, and the earliest of the English herbals were little more than borrowings from the Continental authors and in the case of plants known to the Arabs, thirdhand accounts taken from translations by others from Arabic into Latin. They confuse the *Khubz al ghurab* or nux vomica with the thorn apple or Methel nut (*Datura*) and possibly with other species. The thorn apple is a plant unrelated to nux vomica, and contains the highly poisonous alkaloid atropine, which is chemically unrelated to strychnine. It is widespread around the world and it is not certain where it originated; it is found in England, where the Romans probably introduced it. Like nux vomica, it has toxic seeds, called *Dhutoora* in India. These were used by the Thuggees, the notorious Indian bandits, to intoxicate travellers so that they could be robbed, often with fatal results.[9]

William Turner's *New Herball*, published in 1568,[10] is only able to crib what Matthiolus[11] has to say.

> Matthiolus writeth that the flat nuts like little cheeses, which
> have been sold hitherto for vomiting nuts, are nuttes methel,

and they that have been hitherto used for methel nuts are the right nuces vomicae, that is vomike nuts…the right vomike nuts have little knops upon them like eyes, and the methel nuts have downy or rough skin all over them.

From this it would appear that Turner, or rather Matthiolus, is wrong, according to our modern usage certainly, because the description of the methel nuts that he gives is unmistakably that of nux vomica, for it is these that are like "little cheeses" covered in downy skin. Turner then quotes what Serapio, another Arab writer, says about the vomike nut: "If it be given to the weight of two and two seventh drams, it will kill a man forthwith without any delay; according to the Arabs it engendereth drunkenness, loathsomeness (nausea) and vomiting." These symptoms sound more like those of dhatoora than nux vomica. He says that the seed (he means fruit) of nux vomica is like a lemon, which fits; although this description is cribbed from Avicenna.[12] It is clear that things called "vomiting nuts" were openly available in England in the sixteenth century, although it is not absolutely certain what they were.

There was also confusion about the name "nux vomica." The real reason why the nuts that became firmly known as the vomiting nuts, or nux vomica, do not in fact induce vomiting, is a linguistic misunderstanding. A *vomica* in Latin is an ulcer or abscess, and the name nux vomica, originally thought to be given by Serapio, comes from the use of the powdered nuts in Arab medicine to treat sores.[13] Later writers such as Turner assumed the name came from a power to cause vomiting. But nux vomica is not really an emetic, and much to their distress, many later doctors who inadvertently overdosed their patients found it very difficult to get them to vomit. There was also the practical problem of getting anything down the throat. In 1829, when a 20-year-old woman in Birmingham attempted suicide by taking half an ounce of nux vomica,[14] the physician found that every time he forced her jaws apart to administer an emetic, the paroxysms were so powerful that she bit through the cup. Eventually he resorted to a metal one, which afterwards bore the clear impression of her teeth.

Turner, back in 1568, considered both kinds of nut highly dangerous—too dangerous for human patients to take—and concluded on a sensible note:

...of all that I can gather of these Arabians, the nut Methel
stirreth a man to vomit much more than nux vomica doth,
and that in less quantity. Wherefore, the working of Nux
Methel serveth more the name of the vomiting nut than the
commonly called nut vomike doth. But seeing that it is out
of all doubt that they are very perilous, I will advise all my
friends to use neither of both in their bodies, but to use them
to catch fish, birds and some little beasts therewith; and it
were best to take out of the stomach of such as are eaten
straightaway, and not suffer them to live after they be dosed
or made drunken.[15]

Wise advice so early on, especially for users of his trapping
technique for edible game. A correspondent to *The Times* some
300 years later, "Philo Veritas," related that in Mexico, it was
commonplace to dispose of wolves and buzzards by feeding nux
vomica to an old worn-out mule and leaving the carcass to be
eaten. Not only would the wolves die, but also the buzzards that
fed on their carcasses. At about the same time, on Lord Middleton's estate at Wollaton, Nottingham, one of the gamekeepers
used nux vomica to dispose of some rats and threw the corpses
on the manure heap where they caused the death of all of his
Lordship's chickens.[16] These effects would have been due to unab-
sorbed poison still in the intestines of the dead creature; the flesh
of a poisoned fowl did not kill a dog that ate it, only the innards.
This in itself was a mystery. The poisonous effect of nux vomica
seemed so great, if it could kill without being absorbed, that it
seemed it must emit a "virus" of extraordinary power.[17]

Not long after William Turner, John Gerard in his *The Herbal
or General History of Plants* of 1633 reinforces the view that nux
vomica was by then widely available. But it is clear that the nuts
were just a commodity imported and sold, and that the degree of
botanical knowledge in Gerard's time of the plants from which
they came was still negligible:

Avicen[na] and Serapio make Nux vomica and Nux Methel
to be one, whereas there have been much cavilling; yet the
case is true that the Thorn Apple is Nux Methel. Of the tree
that beareth the fruit that is called in shops Nux vomica and
Nux methel, we have no certain knowledge; some are of the

opinion, that the fruit is the root of an herbe, and not the nut of a tree. These nuts do grow in the desarts of Arabia, and in some places of the East Indies; we have no certaine knowledge of their springing, or time of maturitie.

Of the physicall vertues of the vomitting nuts we think it not necessarie to write, because the danger is great & not to be given inwardly, but mixed with other compositions, and that very curiously by the hands of a faithfull apothecarie. The powder of the Nut mixed with some flesh, and cast unto crowes and other ravenous fowles, doth lull and so dull all their senses at the least, that you may take them with your hands.[18]

Gerard also recommends mixing the powder with some meat or butter and placing it in the garden where cats will "scrape to burie their excrements"— clearly not a pet lover.

What, then, of medicinal uses in this era? For a long period following the writings of Nicholas Praepositus, nux vomica, mentioned only occasionally, was a dangerous substance used mainly for poisoning unwanted animals. We have seen how William Turner and John Gerard cautioned against its use. Most of the remedies put about between the Middle Ages and the eighteenth century were based on European plants and animals and were at least only moderately harmful. If troubled with "Retention of such Humours as foul the Viscera, and stuff the whole Habit with Water and Viscidities," much safer to stick with Dr. W------'s infusion of millipedes.[19]

A superbly illegible manuscript in the Vienna state archives, written in Prague in about 1550 (reproduced in Figure 6), is thought to be the work of one Georg Handschius, and is mostly based on the recipes of the fifteenth century physician Gallus. It gives the ingredients for many medicines and among other things it reports how a nostrum called *Electuarium de Ovo* was enthusiastically endorsed by Habsburg Emperor Maximilian I (1449–1519) as a specific for plague, mania, hydrophobia and poisoning. It can just be made out that a main ingredient is nux vomica; the other ingredients were egg and treacle.[20] The term "treacle," however, originally meant a healing medicine, and mediaeval treacles were complex herbal mixtures, or galenicals. So *Electuarium de Ovo* was a herbal medicine containing various herbs, egg, and nux vomica. None of the sixteenth century physicians who prescribed it could agree on the correct dose, but as all their patients died, the matter was left unresolved.[21] Somewhat later in the same part of the world, confusion about the various bitter plant extracts led to the peddling of nux vomica by a former shoemaker as a cure for malaria, and a mixture of nux vomica and gentian was incorrectly said for many years to be as effective as cinchona bark.[22]

The 1653 edition of Nicholas Culpeper's *Complete Herbal and English Physician* lists more than 300 medicinal plants used in his medicines, but nux vomica is not among them. There is a recipe for *Electuarium de Ovo* sure enough.[23] True to the fiendish complexity of most medicines of that era, it is made by pounding a hard-boiled egg with saffron and 11 other herbs, together with a generous measure of Venice treacle. Given the history of nux vomica's progress from east to west with Venice as a staging post, the latter name sounds promising as a possible nux vomica concoction. But the recipe for his Venice treacle is given clearly enough, and although it contains no fewer than 69 different herbs and venoms, nux vomica is not there, unless misrepresented as his "Dittany of Crete" or his "Indian Leaf," which is unlikely. So it appears that *Electuarium de Ovo* was a name applied to any galenical containing egg and Gallus had the bright idea of pepping it up a little.

Despite the travels of later botanist–explorers, knowledge of nux vomica remained hazy. Even as late as 1859, Alfred Swaine Taylor (of whom we shall hear a great deal later) wrote that the fruit from which the nux vomica seed came was "said to be" the size of a pear.[24] It is easy to forget that during the centuries between 1492 and the construction of the Suez Canal in 1869, distant parts of the Old World, although dimly known to the West since ancient times, were less accessible to European travellers than the Americas. The Philippines, which are a rich source of drug-bearing plants, were not discovered by Magellan until 1521, and not settled by the Spanish until 1565. Any drugs from the Philippines reaching the West until then would have come through the agency of the Arabs, the Chinese or the Japanese, and their origins were fabulous. Although vast quantities of goods were exported from India to Europe, for a scientist such as Taylor to actually go there to see for himself would have been more of a career move than a field trip.

The modern classification of plants is based on the system of the Swedish botanist Linnaeus. In 1753, he proposed the genus *Strychnos*, bringing together a number of trees and shrubs from India and the Far East which seemed to share not only many of their anatomical characteristics such as flower and leaf structure, but also their bitterness and poisonousness. He took the name from the Greek στρυφός, bitter. The term *Strychnos* had in fact been used from Roman times in botanical descriptions, but the plants referred to were what are now classified as members of the potato family (Solanaceae). In using bitterness as one of the characteristics that united the genus *Strychnos*, Linnaeus was presaging the later discipline of chemotaxonomy. This is the attempt to classify plants by their chemical metabolites rather than by their appearance. Modern botanists recognise 196 different *Strychnos* species, distributed almost equally between Asia, Africa and the Americas, although some Asian species are found as far east as Queensland, Australia.[25]

Strychnos nux-vomica became the best known of the *Strychnos* species, but the nuts, wood and bark of several others also reached the West and were traded from time to time. *Strychnos ignatii* grows mostly in the Philippines and like lignum colubrinum had a reputation as a remedy for cholera, fever and snake bites. The beans were known in the Indian bazaars as *Pepita*, a name picked up from the Spanish, but are better known in Europe as Saint Ignatius beans, since legend has it that they were first imported to Europe by the Jesuits, who obtained them as a tribute from the islanders.[26] Saint Ignatius beans also found their way to China, where they were called *Leu-sung-kwo* (Luzon beans). But they were only ever irregularly available in London. They are less starchy and even more poisonous than those of nux vomica. In 1699 the botanist John Ray described how eating them caused giddiness, nausea, faintness and sweating. Poisoning by Saint Ignatius beans is a rarity, but a case was reported in 1900. A 67-year-old woman, the widow of an herbalist, had been in the habit of taking some of her late husband's medicines. One morning when she wanted a pick-me-up, she drank off some extract of Saint Ignatius beans in a glass of hot water before breakfast. After use of the stomach-pump and twenty-four hours of intermittent convulsions, she eventually recovered.[27]

Despite Linnaeus's initial definition of bitterness as one of the determining characteristics of the genus, some species considered to belong to it do not contain any of the poisonous alkaloids. The clearing nut, *Strychnos potatorum*, is said to have edible fruit and, to be a useful emetic in dried form. Its main use was for clearing muddy water. The cut ripe seeds were rubbed on the inside of an earthenware vessel. When cloudy water was poured in, most of the impurities would settle to the bottom of the jar. This use was known not only in India, where the technique was learned by soldiers of the British army, but further east in Burma.[28]

The most notorious member of the genus, growing in tropical India and Burma, was christened *Strychnos nux-vomica* by Linnaeus. It is a deciduous tree up to forty feet high, with a short, thick, crooked trunk giving a very durable close-grained wood. The elliptical leaves are smooth and up to six inches long. The tree bears copious greenish-white, funnel-shaped flowers during the tropical winter season. The flowers, with their unpleasant turmeric-like smell, are succeeded by orange-red spherical fruits, up to three inches in diameter, each containing five of the disc-shaped seeds or nuts embedded in jelly-like white pulp. When fresh, the horny interior of the nuts is white and semitransparent, and has a faint liquorice-like odour. Each seed weighs about 30 grains (two grams). The usual names in India were *caniram* or *kuchila*, but it had many other names in different regions of India. To the British, the usual name was the poison nut tree, but it was sometimes known as the crow fig, and for a time at least in Germany, the seeds were called *Krahenaugen*, or crows' eyes. In the United States they became known as *quaker buttons*, for reasons lost in the mists of time.

The fruit pulp of the nux vomica tree was always said to be virtually non-poisonous, and to be eaten by many birds, as well as by children, with no ill-effect. But in the nineteenth century, the pharmacists Flückiger and Hanbury[29] carried out a careful analysis of all parts of the plant and found even this flesh to be poisonous, as well as all other parts of the tree and even parasitic plants which grew upon it, absorbing its poison. In 1840, Baboo Rajendra Lall Mitra of the Indian hospital service described the case of a stout, athletic English sailor who was admitted into the Calcutta Medical College Hospital suffering from gonorrhoea. His disease was slight, and he would spend his days walking about the hospital compound displaying his pugnacious habits by attacking the servants. One morning, by an "unfortunate mistake," instead of his usual medicine, someone gave him three packets of the powdered leaf of the *kuchila molung*, a parasitic plant (*Viscum monoicum*) that grows on the trunk of the poison-nut tree. On swallowing the first one, he cried out, "I am poisoned!" and from that moment until his death four hours later he was unable to utter another word, each attempt at speech resulting in violent and excrutiatingly painful convulsions, which could also be provoked even by the action of a fly landing on his skin.[30]

It is thought that the nux vomica tree and its relatives propagate by the agency of toucans, monkeys, civet-cats and other animals. These eat the pulp and discard or excrete the indigestible seed, which passes through their bodies without harming them. The elephant bird eats the seeds with impunity.[31] Some authorities stated that nux vomica seeds from Madras and Cochin China usually had a rounded edge, whereas those from some other regions might be more valuable because they had a wavy outline associated with greater potency.[32] But there was always considerable uncertainty about this. In 1883, it was said that the best samples came from Bombay, but even at this late date, the evidence was based on samples bought on the London market and not on observations made in the East.[33] It is possible that some of the nuts reaching London were not those of *Strychnos nux-vomica* at all, but of other closely related plants. There is also a variety of the nux vomica tree, sometimes named as a separate species *Strychnos nux-blanda*, which has sweet-tasting seeds that are not at all poisonous.[34] So there was a great deal of variability in the drugs exported from India, and a lot of scope for passing off inferior samples.

Despite the later scale of the nux vomica trade, even into the twentieth century uncertainty persisted about the correct identification of the various *Strychnos* species. It was not until Dunstan and Short's publication of 1884 that a satisfactory description of nux vomica was published in English,[35] and by that time, as we shall see, the volume of the nuts being imported into London had reached an astonishing 500 tons a year and more. Commerce came first.

Endnotes

1. Stuart, 2004:9.
2. Berger, 1995;7:59.
3. The identity of these two Nicholases is tentative.
4. Lawn, 1963.
5. Clusius, 1567.
6. Translated from the Latin by Philip Powell.
7. Most snakewood was probably the root of *Strychnos colubrina*, a tree closely related to the nux vomica tree *Strychnos nux-vomica* and equally poisonous. It is impossible to identify the exact species from Da Orta's descriptions (he describes two other kinds of similar wood in addition).
8. Pickering, 1879.
9. An Indian newspaper reported in 1833 that a band of 100 thugs was murdering 800 people a month, although this figure may be exaggerated. Their main method of killing was strangulation, but they seem also to have understood poisons. "When Madhajee Scindiah caused seventy Thugs to be executed at Multoun, was he not warned in a dream by [the goddess] Davee that he should release them? And did he not the very day after their execution begin to spit blood? And did he not die within three months?" (Kaye, 1843).
10. Chapman, 1995. Turner was a Cambridge scholar who lived through the turbulent late Tudor period. He was twice imprisoned for preaching without a license or for wearing the wrong vestments, and once had to flee the country.
11. Petrus Andreas Matthiolus. He was born in Siena, Italy, in 1500 but travelled around Europe, and his books on botany and medicine became widely known. He died of the plague in 1577.
12. Abu Ali el-Hosein ben Abdallah ibn Sina (980–1037) is a Persian writer on medicine. His work is considered the culmination of Arab medicine before its decline, and its translations supplanted Galen in the West as the most-consulted source.
13. Terence D. Turner, *Pharmaceutical Journal,* August 18, 1962:151. The nuts also look rather like ulcers, and in traditional medicine were very probably used to treat ulcers for that reason.
14. *Edinburgh Medical and Surgical Journal*, 1829, 31; 446.
15. Chapman, 1995.
16. *British Medical Journal*, May 9, 1857:402.
17. Desportes, 1808.
18. Gerard, 1975. Gerard was even more of a borrower than Turner. This third edition was much revised and improved by Thomas Johnson, and many of his worst mistakes corrected.

19. Coatsworth, 1718. Take four pounds of live millipedes, infuse them cold in eight parts of white wine, then strain for use. By "millipedes," the authors probably meant woodlice, not that it makes any difference.
20. Sigmond, 1837.
21. Husemann, 1857.
22. Desportes, 1808.
23. Culpeper, 1653.
24. Taylor, 1859.
25. Chopra, 1949.
26. Husemann calls this a "Danaëic gift," as well he might. Based on the myth of Acrisius and Danaë, such a gift is one which brings disaster to the recipient (Husemann, 1857).
27. Lancet, 1900, [2]: 486.
28. J.M. Maisch tested the seeds in 1871 and was unable to get a colour reaction for either strychnine or brucine. The seeds, known as Indian gum-nuts, were of so little value that they had arrived in New York as ballast in a ship from the East Indies, and failed to find a buyer, despite their reputation in India of having medicinal properties (John M. Maisch, *American Journal of Pharmacy*, 1871: XLIII).
29. Flückiger and Hanbury, 1879.
30. Chevers, 1870:248.
31. Nadkarni, 1954:1175.
32. *Encyclopedia Britannica*, 9th Ed., 1884.
33. Dunstan, W.R. and Short, F.W., *Pharmaceutical J.*, June 5, 1884 and June 22, 1918.
34. The best modern botanical description of the Strychnaceae is given by Leeuwenberg, 1969.
35. Dunstan and Short, 1884:1; Dunstan and Short, 1918:299.

CHAPTER 3

The Patient Generally Lies on His Back

High where the Fleet ditch descends in sable streams
To wash the sooty Naiads in the Thames
There stands a structure on a rising hill
Where students take their licence out to kill

Samuel Garth, *The Dispensary*, 1699

Although the scientific study of nux vomica, and within a very short time of its active principle strychnine, began in France, the commercial exploitation of the new medicine leaped ahead in Britain.

From about the time of John Gerard, the majority of trade between India and England was controlled by the East India Company, established in 1600 to challenge the monopoly of the Portugese and the Dutch. The eventual scale of the Company's operations was enormous. By 1800, it was employing no fewer than twelve million people in producing and collecting goods for Western markets.

Nux vomica trees grew profusely in the tropical regions of the subcontinent and the nuts could easily be collected from the wild. In 1787, the East India Company founded a 300-acre botanical garden at Sibpur on the banks of the Hooghly, near Calcutta, but true to the ethos of the company, its purpose was entirely commercial and not scientific. The objective was to establish in India the cultivation of valuable plants that grew further east, especially teak from Malaya, and it is very unlikely that nux vomica would have been cultivated there. It was a local plant in abundant supply, and was not strategically important. Many Indian families would have supplemented their minute incomes by sending their children out to pick the Kuchila nuts. In 1870 a whole basketful, enough to poison hundreds of people, could be bought in Rutnagherry (160 miles south of Bombay) for a few pice, a minute sum.[1] Nux vomica was never an expensive commodity, but since it could be obtained in India for virtually nothing, it was a useful extra revenue generator.

The early fortunes of the East India Company were intimately bound up with this trade in drugs. The export of such low-volume commodities (known as fine goods) was a welcome diversion for the company away from gross goods such as textiles and the cheaper spices, and could substantially increase the value of a cargo. But the trade routes were hazardous due to piracy, acts of war and shipwreck. At one point in the eighteenth century instructions were issued not to exceed a certain total value of cargo in case it were lost. This in turn led to complaints that ships were returning from India half empty, causing a problem for the Court of Directors:

> We find ourselves upon this dilemma, if we bring over great quantities of Turmerick, Lacks or other grosse goods, we soone clogg ye markett to that degree that they will not return to us our freight, on ye other hand, if we enlarge our trade altogether in fine goods, which are most profitable, our tonnage will be so little that ye force of our fleets will be too weak for ye treasure of their loading.

The court gave a Mr. Lyon Prager the responsibility of searching out more trade in drugs, and although he was successful, the effect on the company's profitability was minimal.[2]

Another reason for the company's struggle with profitability was smuggling. Their agents and employees were widely scattered and difficult to control. There was moonlighting by employees of their own consignments on the company's or other ships. To try to control potential loss of revenue, British subjects were for some time forbidden to trade with, or even be in, the subcontinent unless authorised by the company, and miscreants were shipped home and prosecuted. In 1788 Timothy Bevan, the Quaker founder of the firm of apothecaries later to become Allen & Hanbury, apologised in a letter for not sending a customer an ounce of musk that he had ordered:

> The India Company have sometime prohibited its importation in their China ships lest its scent should be detrimental to their tea. Hence, all that came was smuggled. This to me was unpleasant who am careful not to meddle with what is fraudulently imported; but I have lately understood that the Company's officers, who are the people that bring it, are sworn to do nothing to the detriment of the Company, [and] it occasions a sort of infringement of the oath which I am not easy to countenance.[3]

Other druggists were not so fastidious. Late the following year, the sailing ships *Gloire d'Anvers* and *Concordia,* under Captain Zimmermann, arrived in the English Channel sailing under the Danish flag of convenience. They were laden with Surat cotton, camphor and other Arabian drugs. The ostensible owner of the ships and their contents was the Vicomte Edouard de Walckiers of Brussels, but the true owners were almost certainly English. In dead of night, the cargoes were offloaded onto smaller vessels lying off the Isle of Wight and made their way to London where they were sold on the black market.[4] We will see later what some of the consequences were of this amazingly porous market in such dangerous materials.

The Kuchila nuts, or nux vomica, arrived at the London docks in the holds of the East India Company vessels. Before cargoes even reached the warehouses, theft was extensive. Colquhoun in the 1790s reckoned that one in four of the port workers was corrupt.[5] There were various subspecies of crook: river pirates stole from the deck cargoes of ships lying at anchor; "heavy horsemen" stole from the holds during daylight hours; "light horsemen" stole from the holds at night; and "scuffle hunters" stole from the quays. Bogus rat catchers would introduce their own rats onto ships coming up the river, so that they would have an excuse to go onto the ships and steal from them. In August 1793 one East India ship was plundered of a large case of pepper, many bags of rice, two sides of beef, and 60-dozen bottles of drink, all in one night.[6]

Until the later development of enclosed docks by the company's successor the Port of London Authority, all cargoes had by law to be warehoused within the city of London. In due course the company developed two town warehouse complexes. One was the Cutler Street warehouse for the secure storage of valuable commodities. This was on a five-acre site leading off Middlesex Street, the famous Petticoat Lane market on the eastern fringe of the city, where folklore had it that it was common to lose your watch at one end and be offered it back for sale at the other. The cynical may be tempted to speculate how much the proximity of Cutler Street may have contributed to the growth of this extensive street market, for inside was 638,820 square feet of some of the most valuable commodities of the eighteenth and nineteenth centuries: carpets, Chinese and Japanese porcelain, cigars, spices, and drugs.[7]

Gums, barks, roots, nuts and leaves arrived at the warehouses in bags, bales and cases. Each package received was examined for damage in transit, classified, sorted for quality and sampled. The labourers developed enormous expertise. It was said that the Cutler Street carpet warehousemen knew more about the origin, age and knotting techniques of oriental carpets than the curators at the Victoria and Albert Museum, who often went to them for advice. Much later, John Masefield, the poet laureate, was to write some exceedingly bad verse after a visit there in 1914:

> You showed me nutmegs and nutmeg husks
> Ostrich feathers and elephant tusks
> Hundreds of tons of costly tea
> Packed in wood by the Cingalee
> And a myriad of drugs which disagree

Nux vomica is listed in an 1888 goods schedule for the East India Company's Fenchurch Street town warehouse, along with about 250 other items, some of them drugs; aloes, ipecacuanha, and various others.[8] The fact that these were not warehoused at Cutler Street shows that they were considered of secondary value and were not carefully guarded. Later on, a specialist drugs warehouse was built at Pennington Street in Shadwell. Old photographs show a dingy, utilitarian building. The area was heavily bombed in the Second World War and no trace of the warehouse remains. Pennington Street is now best known as the address of *The Times* newspaper, which in the 1980s relocated there into a gated glass-walled complex whose security arrangements must greatly exceed in effectiveness any that prevailed at the drugs warehouse.

As the drug industry became progressively more industrialised, improved procedures for handling and controlling dangerous drugs were reflected in the physical layout of plants such as Allen & Hanbury's warehouse in Bethnal Green, London.[9] By 1927 the analytical laboratory there was analysing 8,000 samples a year of crude drugs, chemicals, roots and other substances. Floor D, the floor on which dry drugs were packed, was dominated by long rows of bales, boxes and canisters containing roots, barks, gums and powders. Part of the floor was occupied by the poison cage of stout wire, in which all dry poisons were stored. Within this again was another cage containing the Dangerous Drugs section, where narcotics such as opium were packed by girls under the supervision of a full-time clerk able to track the slightest discrepancy in the weights issued and packed. This highest level of security was reserved for the highly addictive morphine and its relatives. Nux vomica was not among these. It was in the outer cage.

There were several nominally distinct professions involved in dosing the gullible London public during the era of the East India Company. Drugs from abroad, such as nux vomica, were for many years imported by the Grocers, who were originally called the Grocers, Spicers and Pepperers.[10] This was an ancient and prestigious city company that made huge sums from the import trade. The apothecaries, or drug specialists, not having their own company, were for many years kept under the grocers' thumbs. From the earliest times the Grocers' Company punished errant members for adulteration and misrepresentation. An official called the Garbler of Spices was allowed to enter any shop or warehouse to garble, that is check, the purity of spices or drugs, and it is easy to see how this was used as a device for keeping the apothecaries in line.

The physicians obtained their royal charter in 1518 and at first prepared their own medicines, but before long they realised that the apothecaries could undermine them by selling remedies directly to the public. During an outbreak of plague in 1562, the physicians lost a lot of credibility by leaving town. On their return they found it necessary to launch a propaganda war accusing the apothecaries of selling dangerous remedies, and began to agitate for taking over their supervision from the grocers. In response, the grocers increased their garbling activity and seized from the apothecaries and burnt a lot of "noughtye" and corrupt drugs. The physicians then stole a march on the apothecaries by opening a botanical garden, with John Gerard as curator. In 1588 the apothecaries petitioned Queen Elizabeth for a monopoly on the supply of drugs, but were unsuccessful. In response to continuing pressure, in 1607 the grocers modified their charter to recognise the apothecaries as a separate section, but this still did not give them any power. So in 1610 the apothecaries, under the leadership of a French Protestant residing at Blackfriars, Gideon de Laune, petitioned King James I for their own company. They pointed out that many unskilled people were supplying drugs to the public, and that the grocers were incapable of knowing good drugs from bad. Some of the drug sellers preferred the status of staying with the Grocers to joining the breakaways, and tried to undermine the independence faction by describing them to the king as tobacco sellers, capitalising on his well-known aversion to the weed. But they were unsuccessful, and the Apothecaries Company secured its charter in 1614. The king, a dour Scot not too enamoured of overfed city traders, took the part of the apothecaries:

> Another Grievance of mine is that you have condemned the patents of the apothecaries in London. I myself did devise that Corporation, and do allow it. The Grocers who complain of it are but Merchants...they bring home rotten wares from the Indies, Percia and Greece and therewith through mixtures make waters and sell such as belong to the Apothecaries, and think no man must control them, because they are not Apothecaries...[11]

The apothecaries, however, although now nominally a separate profession, had enlisted the support of the physicians in their fight for independence. They now found themselves treated as assistants, with the physicians demanding the right to inspect their premises just as the grocers had done. In 1650 the physicians obtained an order in council that no apothecary was to sell any "poyson, drugge or medicine" except on prescription from a physician. The apothecaries counterargued that if they were allowed to make drugs but not sell them, then the same reasoning would mean that the cutlers would not be able to sell the knives that they made, because these were just as dangerous as medicines. The succeeding two centuries and more were marked by a continuation of this jostling for power, each group trying to exclude others from handling the highly profitable drugs and poisons. Their various efforts at control ultimately foundered on the inability to define what, if anything, ought to be controlled.

Parliament showed an enormous and protracted unwillingness to legislate to stop people dosing themselves and other people with whatever they wished. In the eighteenth and early nineteenth centuries, the domestic medicine cupboard usually contained a variety of dangerous shop-bought medicines including nux vomica. It was widely believed that you should dose yourself on principle at least once a year, and that the more unpleasant the disease, the stronger the medicine should be. Unlike the physicians, the apothecaries were forbidden to charge for advice, only for the medicine they dispensed. The expectation that a visiting practitioner would always prescribe a medicine became deeply ingrained in the minds of the British public. Whatever their ignorance of medicine, the apothecaries became immensely successful financially. The coach-keeping apothecary, living in style by selling rare drugs, became a well-known city figure. So much envy was generated that in 1707 there was a petition to wind them up,[12] and in 1748 parliament refused to confirm their monopoly. But in 1776, no less a figure than Adam Smith, the author of the *Wealth of Nations*, was still able to write that "apothecaries' profit has become a byword, denoting something uncommonly extravagant." At about this period, apothecary Daniel Graham lived in Pall Mall with his wife and four children and had their portrait, surrounded by all the trappings of an aristocratic lifestyle, painted by Hogarth, a painting which today hangs in the National Gallery.[13]

Then there were the chemists, who based their doctrines on a more recent tradition going back to Paracelsus (Theophrastus

Bombast of Hohenheim, 1493–1541). Born in Swabia, in his short life Paracelsus travelled in Africa, Arabia, Italy, Scotland, Ireland, Cornwall, Moscow (where he was taken prisoner by the Tartars), the Baltic states, Hungary, Crete, Egypt, Ethiopia and Arabia again. Although he was an alchemist and his theories did not strictly speaking advance science very much, he was a famous iconoclast. He publicly burned the works of Galen and Avicenna and persuaded the alchemists to abandon their search for gold and look for useful medicines. He maintained that every disease should have a cure somewhere in the natural world, but not necessarily the plant world. Within the spread of those who called themselves "chimysts" were natural philosophers, such as Robert Boyle, gentleman investigator and the first to define the term chemical element in 1661; we would call them pure scientists. Others were effectively pharmaceutical tradesmen, and there were all gradations in between.

The public, particularly outside London, could buy dangerous drugs from totally unregulated and untrained people calling themselves chemists or druggists. Chemists believed in dosing people with synthetic chemicals such as mercurial ointments to cure syphilis and tartar emetic to produce vomiting.

During the era of Turner and Gerard, most medicines supplied by apothecaries were elixirs; complicated concoctions of many different herbs aimed at curing more or less everything, taking many hours to prepare and correspondingly expensive. A particular kind of elixir was the mithridatum, named after Mithridates the ancient Anatolian king, who believed in taking small amounts of every poison mixed together to protect him from their effects.[14] This reflected the belief of many doctors even into the nineteenth century that all diseases, except mental diseases or those caused by debauchery or other excess, were forms of poisoning.[15] Such complex medicines were known as "galenicals," as opposed to "simples," which were the extracts of a single herb. Part of the training of apprentice apothecaries in the eighteenth century was all-day plant-collecting, or simpling expeditions starting at Apothecaries Hall at five o'clock in the morning. As London expanded, the amount of walking necessary to reach open countryside became excessive and many of the apprentices only turned up for the free meal afterwards. The society then purchased a poorly constructed barge to convey the apprentices up- or downriver, including trips to their Chelsea Physic Garden which they had started in 1673 and later replanned on Linnaean principles as a result of a (somewhat acrimonious) visit from the great Linnaeus himself. The society also built a laboratory for compounding galenicals in 1623, and a chemical laboratory somewhat later, but as these were sited at Apothecaries Hall at Blackfriars in the middle of London, there were constant fires and complaints from the neighbours about the stink.[16]

A delightful snapshot of the trades of chemist and apothecary is given by R. Campbell, Esq.,[17] strolling the streets of London in 1747. He is in no doubt about which group of tradesmen he prefers. The chymist, he asserts, must have solid judgment, unwearying patience and acute observation. He must be a man of honour and conscience, because he has many opportunities for foisting fictious compositions upon the public, and thus by ignorance, greediness and villany depriving the patient of his life and the physician of his reputation. He must be of robust constitution and a master of Latin and German, because apart from Boyle, who is a windbag, "we have few else in the English Tongue that make any Figure; therefore the young Chymist must have recourse to Foreigners," especially the Germans, who are the best chemists in Europe. But he has to admit that the practical chemists who go under that name in London are "far from being adepts in this Study." They are "seldom employed about any part of their Branch which does not immediately depend upon the Practice of Physic"; in other words, they are virtually indistinguishable from the apothecaries and druggists, whom they are able to undercut because they make up their own chemical ingredients.

The apothecary, however, is a being of a lesser stamp:

> His Knowledge, by his Profession, is confined to the names of Drugs...he must only know that Rhubarb is not Jesuit's Bark; that Oil is not Salt; and that Vinegar is not Spirit. He must be able to call all the Army of Poisons by their proper Heathenish Names, and to pound them, boil them, and mix them into their proper Companies, such as Pills, Bolus's, Linctus's, Electuaries, Syrups, Emulsions, Juleps, etc. He must understand the Physical Cabala, the mysterious Character of an

unintelligible Doctor's Scrawl....This is a mere Apothecary, a Creature that requires very little Brains...

Could it be that by any chance the author is motivated by unworthy envious feelings? For, he tells us, the cost of setting up as an apothecary is a mere ten or twenty pounds, which will buy all necessary gallipots and vials, and fill them with enough drugs to poison the whole island. Yet his profits are "unconceivable"— 500% is the least he receives. Not content with this extortion, he has no sooner set up shop than he commences to doctor people, only calling in the physician to be present at the death of the patient, or to justify the enormous bills he has run up for treating ailments like venereal disease, for most clients find that "their Mistresses have only clapped them, but Doctor Apothecary has poxed them."

Things changed little with the passage of time. *Punch*, in 1849, was if anything even more dismissive of the apothecaries than Campbell had been a century before. Incensed by their refusal to license graduates of the newly established College of Chemistry in Liverpool, Mr. Punch fulminated:

> The professors' chemical attainments, I perceive, are attested by the best chemists in the world, with LIEBIG at their head....I express my wonder at your impudence in withholding your recognition of it....Is not your Company a society of medicine-vendors, dealing also in pepper and having a shop for the sale of its wares at the top of Union Street, Blackfriars? What are you but a fraternity of spurious physicians, who originally picked up a smattering of medicine by making up prescriptions?....Understand, worshipful Sirs, I do not object to your trade, I only wish you would stick to it, and not aspire to dictate to a liberal profession...[18]

What all these factions had in common was a desire to control and disenfranchise those who came at the bottom of this ecosystem, the quacks. But the definition of a quack was a very hazy one. There is little in principle to distinguish Robert James, the inventor of the James's powders that killed the Irish writer and dramatist Oliver Goldsmith, and the Beecham who invented Beecham's pills and went on to found one of the leading British pharmaceutical companies. In the following century, *Punch* put the case with appropriate sarcasm:

> The current medical bill is intended to put down quacks but we can see no difference between the vendor of a patent pill for every disease, and the family doctor, who continues to send medicine which he knows will do no good, for the mere purpose of running up a bill in which "MIXTURE-MASTER JOHN" shall be repeated some twenty or thirty times, at three shillings a bottle, and "Pill and Draught-MISS ELIZA" shall run through a page and a half of foolscap at eighteen pence per item....who address[es] childish enquiries to the baby as to its little "tonguey pungy" and promise[s] to send the unhappy infant some "nicey picey" for which they intend to "chargey pargey" to a tremendous "summy pummy."[19]

The answer to all of this was legislation, examinations and the enforcement of professional standards; but of course this could only go hand in hand with genuine advances in medical knowledge. Accustomed as we are to the bodies, such as the General Medical Council, that were set up as a result of tidal waves of reform that crashed through Victorian society, we can fail to appreciate just how slapdash things were during the preceding decades. In fact it was the apothecaries in 1815, given control over all general medical practice in the country, who led the way by introducing written examinations in 1839; before even the Civil Service. The concept of having to submit to a written examination rather than an apprenticeship before practising a dangerous trade was slow to catch on. Alfred Westwood, the son of a freeman of the Apothecaries' Company, pleaded to be allowed to practise, despite not having taken the examinations. He had studied, he said, in Paris and elsewhere, and had complied with the spirit of the Act of Parliament by helping in his father's shop. The company rejected his plea, but not apparently out of hand, only after considerable soul-searching.[20]

Despite these initiatives, a professional qualification for someone supplying powerful drugs to the public remained an option, not a requirement. By the mid-19th century there were seven hundred shops in London, more or less equally divided between two sets of identical-looking establishments, each with toothbrushes and large bottles of coloured water in the window. But one set was run by qualified licentiates, and the other by totally unqualified druggists who may not necessarily have known the difference between fungal disease and cancer, or between nutmeg and nux vomica.

The lethargic Apothecaries Company was always vulnerable to falling between two stools. Many of its members wanted to practise medicine. After the Medical Act of 1858 they eventually joined with the physicians to form the modern medical profession, although the snobbish attitude of the physicians, now become Harley Street doctors, to the general practitioners, former apothecaries, lasted well into the twentieth century. (Surely it is extinct today?) Those apothecaries more interested in making and selling drugs formed the Pharmaceutical Society in 1841 and under the leadership of the great Jacob Bell gradually began to impose professional standards on the hitherto unregulated druggists and tradesman–chemists. The academic chemists formed the Chemical Society of London at the same time, though not without the persistence even to the present day of a certain amount of ill-feeling over the pharmacists continuing to appropriate the term "chemist" in the minds of the general public.

Until the foundation of the Pharmaceutical Society and beyond, even those clutching a certificate remained dangerous. The quality of scientific education meted out to them was atrocious. One recipient of the necessary course of instruction was a certain William Palmer from Rugeley in Staffordshire, who on April 15, 1847 signed the register of the Apothecaries Company. No doubt he was clutching his copy of John Steggall's *Manual for Students Preparing for Examination at Apothecaries Hall*, a farrago of medieval alchemy and ill-described medicine, the greater part of which could have been written at any point in the previous two hundred years. In this book, students are coached to answer questions by rote, such as "How is strychnine procured? What are its qualities? What ingredients are used?" A medical diagnosis according to Steggall: congestion of the brain. "There is frequently constipation, and the patient generally lies on his back." His recipe for *Hydrargyrum con creta*: rub together mercury and chalk. "This is a mild alterative, most suited for diseases of children." (Mercury is horribly and cumulatively poisonous, especially for children.) His recipe for *Confectio Sennae* or Lenitive

Electuary reads as follows, *in toto*: "This preparation requires much care and patience in making, and as, from the number of ingredients ordered, some may be omitted without easy detection, it is seldom found in the shops as good as it should be."[21] How useful!

It is unfair of course to judge Steggall by the standards of the present day. But comparison with good material of the same era, for example Christison's *A Treatise on Poisons*[22] remains astonishing. Christison is recognisably a modern scientist. Steggall is at best an alchemist, at worst a buffoon; yet Christison's book was published as early as 1829, while Steggall, who in that year was only just over thirty, continued to bring out new editions for another three decades. Prospective students were able to read in the back of his books that "Dr. Steggall continues to assist Gentlemen in their studies by Private Instruction. The hours devoted to this purpose are, in the Morning, from Eight to Nine and Ten to One; and from Three to Five and Six to Nine O'Clock p.m." The terms for examinations to Apothecaries' Hall were five guineas, for the Royal College of Surgeons five guineas, and for tuition in Latin authors, the same. Those desirous of becoming seriously incompetent could pay twelve guineas for all three, and might also wish to learn that "Dr. Steggall accommodates two resident pupils in his house where they have access to a Museum of Anatomy, Materia Medica etc. and possess the means of following up every branch of their studies in the most efficient manner."

These, then, were the qualified apothecaries. Into their hands the paying public of the first half of the nineteenth century entrusted themselves, unless, of course, they preferred someone unqualified.

Endnotes

1. Chevers, 1870; 241.
2. Furber, 1948; 291–292.
3. Cripps, 1927. Allen & Hanbury eventually became a part of multinational drug giant GlaxoSmithKline.
4. Furber, 1948; 137–138.
5. Colqhoun, 1800.
6. Ibid.
7. Owen, 1927
8. Exhibit at London Docklands Museum, 2003.
9. Cripps, 1927.

10. Rees, 1923: Copeman, 1967.
11. Barrett, 1905.
12. Pitt, 1707.
13. In the early eighteenth century, an apothecary supplied to Mr. Dalby of Ludgate Hill, the following items in *one day*; An emulsion; A mucilage; Gelly of hartshorn; Plaster to dress blister; An emollient glister; An ivory pipe armed; A cordial bolus; The same again; A cordial draught; The same again; Another bolus; Another draught; A glass of cordial spirits; Blistering plaster to the arms; The same to the wrists; Two boluses again; Two draughts again; Another emulsion; and Another pearl julep, at a total cost of £6/6s (Bell, 1841:2).
14. Stuart, 2004: 112.
15. See for example Fouquier, 1802.
16. Barrett, 1905.
17. Campbell, 1747.
18. What *Punch* did not mention was that the Liverpool College of Chemistry had been founded in the stables behind his house by the pretentious and deeply suspect Sheridan Muspratt (Brock, 1997: 125).
19. *Punch*, 12 April 1856: 141.
20. Barrett, 1905.
21. Steggall, 1838.
22. Christison, 1829.

CHAPTER 4

M. Vauquelin's Lack of Fame

Fame is a bee
It has a song
It has a sting
Ah, too, it has a wing

Emily Dickinson, *Poems*, c. 1862–1886

After Lavoisier's execution, his legacy in the field of chemistry passed to a number of somewhat less talented figures, such as Nicholas Vauquelin, who was 21 years old when Lavoisier was executed in 1794. A native of Calvados, Vauquelin was of humble origins and did not speak any foreign languages, but in 1797 he discovered the element chromium, and gained high status in the new post-revolutionary society.[1]

Vauquelin seems to have been a quick thinker.[2] The close relationship between medicine and chemistry, its subdiscipline, at this time is emphasized by the fact that in order to succeed his colleague and patron Fourcroy as professor of chemistry after he died in 1809, he was obliged to have a doctorate in medicine which, fending off his rivals, he obtained in double-quick time by some studies of brain tissue. Thus, like Fouquier and many of the other French doctors of the era, he was intensely interested in the structure and functions of the nervous system. Like Fouquier, he became interested in plants and plant medicines, and there were strong echoes of the other preoccupation of contemporary physicians, the blood, in the way in which he thought about

47

plants. He considered their sap to be analogous to blood, and thought that a careful analysis of sap at every stage of a plant's growth would unlock the secrets of its growth and development, another highly simplistic idea that was to turn out to have little real value.

One of the plant products that may have influenced his belief in the importance of sap had been brought back not long before from the other side of the globe. As well as funding Vauquelin and other Parisian professors, the French government sponsored many foreign expeditions. In 1801, two ships under the command of Captain Baudin sailed on a major voyage of exploration to the Southern Hemisphere. More than twenty scientists and artists left France, but only half returned; some died, others left because of political disputes with the captain, while others fell ill and were left behind at various ports of call. The latter fate attended the chief naturalist, Leschenault,[3] who became sick when the expedition called at Timor for the second time in June 1803, and had to remain there for the time being. While convalescing he noticed that when the locals went hunting, they dipped their arrows into a substance that was clearly immensely poisonous. He persuaded one of them to show him how it was made. The man took scrapings from the root of a local plant, then boiled them in water and concentrated the extract to a thick syrup. He showed Leschenault how, if a sharp sliver of bamboo was dipped in the mixture then shot into a chicken, the bird would die within two minutes.[4]

The expedition brought back more than 200,000 specimens to France, including live emus and kangaroos as well as exotic plants.[5] Also present in the ships' holds were Leschenault's syrup as well as some leaves and branches of the unidentified tree. In 1809, François Magendie and a colleague Raffeneau, using Leschenault's syrup, demonstrated the poisoned arrow experiment on a dog in front of members of the *Academie des Sciences*, to great acclaim.[6] Magendie was also able to show that if the hind leg of a live dog were cut off, leaving it attached only by the main artery and vein, injecting the poison into the near-amputated appendage killed the dog as rapidly as if it had been whole. This and similar unsavoury experiments proved that the poison functioned by initially circulating in the bloodstream. The botanist Jussieu was able to identify the plant material brought back by Leschenault as being from a *Strychnos* species. It began to appear probable to the French investigators that the various plants of the genus given this name by Linnaeus in the previous century all contained the same poison, or very closely similar poisonous principles.

As yet, though, there was no real consensus that a plant with powerful drug properties, like cinchona or nux vomica, might exert activity through the medium of well-defined chemical constituents that it might produce. This was despite some brilliant work by the Swedish chemist Scheele the previous century, who had isolated various natural products, such as citric and tartaric acids, in pure form from plants. Perhaps because Scheele was Swedish, the Parisian chemists tended to think his research irrelevant to the mysterious activities of the drug-bearing plants, which seemed to be mediated through the whole tissue. Vauquelin and others in different parts of Europe did make some attempts to purify the active principle of cinchona bark, but failed, partly because the techniques were lacking and partly because it was a half-hearted exercise; their failure reinforcing the view that a plant's medicinal properties were associated with the harmony of its constituent principles, and would never be properly understood.

There were others, though, who were taking a more reductionist view in the true revolutionary spirit. Already in 1808, Eugene Desportes had written a short thesis on the nux vomica plant and the deadly powder that could be made from its seeds, and described how he and a colleague Braconnot had carried out some rather primitive experiments aimed at finding out what the active principle of the powder might be. They obtained from it something which they called *Le principe jaune amer* (bitter yellow substance). This was a partially purified extract of the plant, but not the pure alkaloid; it resembled the deadly syrup of the Upas Tiute in its poisonousness and seemed therefore to be closely related to it in some a yet not fully understood way. And in Gemany, a few years previously, the apothecary Friedrich Wilhelm Serteurner had isolated morphine (strictly speaking, one of its salts) from opium, and shown that the crystals he had obtained were far more potent than the crude extract of poppies. This discovery was initially overlooked by the Parisian scientists, but when the details were republished a few years later in a different journal, this time it was noticed by Vauquelin and others in Paris, who, modifying their earlier theory that only the whole plant could produce physiological effects, began to replicate Serteuner's type of experiment on cinchona bark, nux vomica and other highly active plant drugs. Vauquelin was probably the first to propose that plants might contain a whole range of highly active substances capable of forming salts with acids. At the time he called them "organic alkalis," or "acid-fixing substances"; the name "alkaloid" came later.

The concept of a research group in the modern sense of a team of people working towards the same set of goals under the direction of an established scientist did not yet exist. Paris remained a rabbit-warren of small scattered laboratories, many attached to the various hospitals, and chemistry was taught by demonstration, as was every other subdiscipline of medicine. A particularly promising student, having learned his craft by listening to lectures in vast auditoria alongside hundreds of his contemporaries, and by watching carefully prepared demonstrations by the lecturer, might then, if he were lucky, be taken on as the professor's personal assistant if the funds were available.

Thus in the same years that Fouquier launched into his experiments on the medical uses of nux vomica, Vauquelin and his students and associates began to repeat and extend the experiments of Desportes and Braconnot. It was two younger and more talented colleagues who made the breakthrough.

Another associate of Lavoisier had been the apothecary Bertrand Pelletier, who although dying at the age of thirty-six had filled several posts in the revolutionary government. His son, Pierre-Joseph Pelletier, born in Paris in 1788, in turn became a chemist and an associate of Vauquelin. Pelletier the Younger would eventually become one of France's greatest scientists. He was the first to isolate chlorophyll from green plants and quinine from the Jesuit bark and went on to establish a considerable business in producing this antimalarial drug and other alkaloids on a commercial scale.

Pelletier and another colleague, Joseph-Bienaimé Caventou, were able to get their hands not only on nux vomica, but also on the elusive Saint Ignatius beans and on *Bois du Couleuvre*. Pelletier and Caventou extracted grated St. Ignatius beans with ether and evaporated the solvent to obtain something greenish in colour and of the consistency of a thick oil or butter. Initially they regarded this as being the pure essence of the poisonous plant, but they then realized that the oil resembled in all of its poisonous properties the bitter yellow substance that they obtained from nux vomica using Desportes and Braconnot's method (and also Leschenault's gummy extract brought back to Paris more than a decade before). They did not believe, they said, that the physically different extracts from the three different plants could not have the same active principle. Eventually, after treating their extract of Saint Ignatius beans with alkali, they obtained an abundant white crystalline precipitate, extremely bitter to the taste, that they recognised as the true active principle. In these beautiful but sinister crystals resided all of the poisonous properties of the plant itself.[7]

There were several important aspects of Pelletier and Caventou's work. They and others obtained the same substance from all three *Strychnos* plants, confirming Linnaeus's views on the close relationship between the members of the genus and reinforcing the academy's interest in using medicinal properties as a way of establishing the relationships between plants. They recognised strychnine as the second member after morphine of the new class of alkaloids; compounds from plants that form salts with acids, and have powerful effects on the organism. They stated clearly, possibly for the first time, the scientific principles of plant drugs. The medicinal effects of a plant, they said, are due to one or more active ingredients; closely related plants have similar active ingredients; and if a member of a group, such as the clearing-nut tree, lacked or nearly lacked it, then it would lack the medicinal properties also. Not least, they produced strychnine in crystalline form. This indicated that they had probably now obtained the pure alkaloid. These guiding principles were only just emerging, and it is notable that when they began their research, Pelletier and Caventou were quite willing to believe that a sticky, greenish mass might be the pure substance.

For the early chemists, producing a substance in crystalline form was a vital step in being able to identify it. One could look at the crystal form under a magnifying glass and measure the temperature at which the crystals melt, hopefully thus distinguishing it from other compounds. Because of the forensic implications, a vast amount of effort over the following decades went into trying to establish reliable methods for identifying strychnine and for assaying it, that is measuring how much was present in a given sample of material, perhaps the organs of someone poisoned by it. Strychnine, as it happens, does not have a sharp melting point; if heated slowly, it begins to melt at about 268 degrees centigrade, but if heated instantaneously to just below 290 degrees, and then more slowly, it melts at this higher temperature. This behaviour indicates that at above 268 degrees it is starting to decompose with the heat, producing impurities. Strychnine is also polymorphic; that is, it can crystallise in six or seven different crystal forms, depending on the temperature and solvent of crystallisation. Polymorphism is quite a common phenomenon, but it was a problem for the forensic workers who studied strychnine. If a sharp melting point was not a characteristic that could be used to identify strychnine, and neither was a unique and recognisable crystal form, then what was? Pelletier and Caventou relied for strychnine's identification on the formation of various salts, and various reactions of the alkaloid with other chemicals such as nitric acid to produce coloured products, reactions which were to assume great significance in the decades to come.

Pelletier and Caventou could hardly wait to study *("Nous nous hâtâmes d'examiner")* the properties of their extraordinary new alkaloid. When dissolved in hot alcohol containing a small quantity of water and allowed to cool, it crystallised in minute four-sided prisms surmounted by four-sided pyramidal faces. The brave investigators found it to be of an appalling bitterness (*"d'une amertume insupportable"*) with an unpleasant aftertaste resembling that of certain metallic salts. The bitterness they found to be detectable at a dilution of anything up to one in four hundred thousand, depending on the sensitivity of different individuals. It had no odour and underwent no change when exposed to the air. It was very sparingly soluble in water but, when treated with acid, it formed stable salts, such as the sulphate and the hydrochloride. All of the salts of the new alkaloid, they found, retain the toxic properties. When treated with dilute nitric acid at low temperature, the nitrate salt was obtained, but with hot dilute nitric acid or with concentrated nitric acid, a chemical change took place and the toxic properties were lost. This loss of toxicity was accompanied by a series of colour changes; firstly an amaranth purple, immediately changing to blood-red which then faded to a pale yellow gradually becoming more intense. Other concentrated acids gave an immediate red colour. They investigated the effect of reacting it not only with nitric acid, but also with other simple chemicals such as tin chloride and sulphur; their theoretical knowledge of what might be going on deserted them, and they were only able to put forward certain simple ideas based on the reactions of inorganic salts. But their efforts represented one of the first attempts to carry out chemical modifications on a natural product.

Like earlier French workers who had experimented with crude nux vomica powder, Pelletier and Caventou, in collaboration with Magendie, carried out some animal experiments on their pure alkaloid. Half a grain introduced into the mouth of a rabbit caused convulsions and the animal died within five minutes. Having discovered such a deadly substance, and finding that the toxic properties cannot be neutralised by simple means such as acid treatment, they argued that the only treatments for poisoning are firstly mechanical means such as stomach pumping to remove the poison before it can be absorbed, and secondly minimising the spasms. To this end, they carried out various experiments in which they simultaneously fed rabbits with a lethal dose of strychnine accompanied by increasing amounts of opium. They had some slight success; the rabbits at least survived longer, and one of them recovered, but they did not consider the experiments successful enough to claim the discovery of an antidote.

There was much interest subsequently in finding an antidote to strychnine poisoning. M. Donne of Paris claimed that strychnine could be destroyed in the stomach by prompt administration of chlorine water. This is a fearful antidote, as any chemist will know; and since it had to be given promptly, even if it worked it offered no advantage over the stomach pumping. Inducing vomiting in those poisoned by strychnine is highly unreliable, for later doctors were to report that even injecting the extremely powerful emetic apomorphine sometimes had no effect. Pelletier and Caventou were correct in using soporifics to counter the spasms;

later, doctors would use chloral hydrate, barbiturates, and later still, massive doses of benzodiazepine sedatives such as valium. By experiments with frogs, a later physiologist de Meyrignac[8] showed that the highly poisonous alkaloid nicotine has the contrary action to strychnine, and is also an antidote, and victims of accidental poisoning were often advised to smoke furiously.[9]

Another approach was to administer something that would either absorb the poison, charcoal being well known for its absorptive properties, or to line the stomach and prevent strychnine's assimilation. To a pharmacist, M. Touery, should go an award for outstanding recklessness in the cause of science; in 1831, in front of the luminaries of the French Academy, he swallowed ten times the lethal dose of strychnine mixed with charcoal, and survived.[10] This is especially noteworthy given that others had claimed that charcoal had little or no effectiveness! A large quantity of lard poured down the throat to line the stomach was often recommended, but American physician Dr. William Hammond thought this was rubbish:

> I gave two grains of the poison [strychnine] to one dog, without the antidote, and two to another, with the addition of a pint and a half of melted lard. The best of the joke is that the latter died in four hours, and the former, miserable worthless cur, who doubtless was too mean to die, is still running about in the finest possible state of health. So much for lard. We are of opinion here that strychnia is quite harmless, unless lard is indulged in.[11]

François Magendie recognised that strychnine acts on the spinal cord. The realisation that the nerves running to and from the spinal column from the various parts of the body are responsible for the sensory and motor functions is said to date back to Galen. The story has it that the sophist Pausanias experienced a loss of sensation in two of his fingers. Treatment of the fingers by various physicians worsened the condition. When Galen was consulted, he elicited the information that on his journey to Rome, Pausanias had been thrown out of his carriage and had suffered a small injury to his spine. Galen then cured the illness in the fingers by treating the patient's back. Vivisection experiments by Sir Charles Bell and by French physiologists, some of them considered shocking by British writers, showed clearly that there were two sets of nerves leading to the spinal column, one associated with sensation, one with movement. Clearly, strychnine interfered powerfully with the latter set. A decapitated animal continued to show the spasms after poisoning with it.

Magendie was also the first to pose two key questions that were to occupy, and mislead, the medical world for many years to come. Firstly, is strychnine of any use in treating disease? Magendie says yes. He has administered an alcoholic solution to a 70-year-old patient suffering from muscular debility as a result of a cerebral malady, presumably a stroke, and after eight hours had obtained *"Une amelioration remarquable dans ses forces musculaires."* Secondly, are the symptoms the same as, or different from, those of the natural disease tetanus? Magendie says they are similar *("J'ai obtenu sur ce malade des effets non-équivoques de secousses tétaniques")* and can hardly have imagined how this point would be argued over in such detail some three decades later in an English courtroom.

There are no definite rules for naming natural products even today. This is in contrast to the taxonomic protocols for naming biological species, which are so strict that a name, once allocated, cannot be changed even if a mistake has been made. When Linnaeus unwrapped a parcel of plants sent to him by a contact in Peru and called a new plant he found inside *Scilla peruviana*, the name was established and had to stay, even after he found that he had mixed up the parcels; *Scilla peruviana* grows in Spain, not in Peru.

In 1818, Magendie suggested the name *Tetanine* for the new alkaloid which his colleagues had isolated from nux vomica and its relatives, but the grant of name rested with its discoverers. Pelletier and Caventou instead proposed *Vauqueline* in honour of M. Vauquelin, but were dissuaded by the commissaires of the French Academy of Sciences who told them that *"Un nom chéri ne pouvait être appliqué à un principe malfaisant"*; a cherished name should not be applied to a harmful principle. So they settled for *Strychnine*. Alas for poor Vauquelin, who passed more or less unregarded into history, lacking the notoriety of any association with Rugeley, Staffordshire, or London's Lambeth Road.

Endnotes

1. Pariset, 1833.
2. During the turmoil of the revolution, a royalist Swiss guard, fleeing for his life, ran down a street and through a door into Vauquelin's laboratory where he was alone with two women (Fourcroy's two sisters). They burnt the soldier's uniform, shaved off his moustache, blackened his hands and face, and passed him off as Vauquelin's laboratory assistant (Pariset, 1833).
3. Jean Baptiste Claude Théodore Leschenault de la Tour (1773–1826).
4. This very rapid-acting arrow poison would also have contained the curare alkaloids, which are chemically closely related to strychnine but are not active when given by mouth.
5. The emus and kangaroos ended up in Josephine Bonaparte's gardens at Malmaison.
6. Bo R. Holmstedt and Jan G. Bruhn, *Ethnopharmacology; A Challenge*, in Schultes, 1995; 338–348.
7. Pelletier and Caventou, 1819.
8. de Meyrignac, 1856.
9. It might be thought that if nicotine is an antidote to strychnine poisoning, then strychnine must reciprocally be an antidote to poisoning by nicotine, thus justifying the belief in nux vomica by da Orta's mongooses. Biochemistry is not as simple as that. According to modern medicine, there are apparently no effective antidotes to nicotine poisoning.
10. Fenton, 2002: 6. Many books, journals, and websites report this anecdote, but at the time of writing I have been unable to locate any original account of Touery actually giving this reputed demonstration. Pierre-Fleurus Touery certainly existed, but his career is poorly documented. An account of his work on charcoal was given in 1902 by two authors who included his grandson, but they make no mention of him experimenting on himself (Secheyron and Daunic, 1903). It may be a case of inaccurate "Chinese whispers" in the literature.
11. Hammond, William H., *American Journal of Medical Science*, cited at Chevers, 1870; 245. A young girl who had swallowed an unknown amount of strychnine was treated by forcing at least a pound of a "filthy mixture" of charcoal and lard into her stomach, followed by a large amount of infusion of tobacco leaves. She recovered after dreadful spasms. The doctor thought the recovery owed more to the nicotine than to the lard or the charcoal.

CHAPTER 5

Perfidious Dutchmen Bark up the Wrong Tree

"It is very fortunate," she observed, "that such substances could only be prepared by chemists; otherwise, all the world would be poisoning each other."

"By chemists and persons who have a taste for chemistry," said Monte Cristo carelessly.

Alexandre Dumas, *The Count of Monte Cristo*

The history of strychnine is intimately bound up with that of a second alkaloid, brucine, which occurs with it in plants of the *Strychnos* genus. The story of how this second alkaloid, chemically closely related to strychnine, came to be discovered, will explain some of the subsequent events in strychnine's history.

Toward the end of the eighteenth century doctors in Trinidad began to recommend a new drug for use against "intermittent fevers" (malaria). It was the bark of an unidentified tree, and the physicians had learned of its use from Catalonian monks resident on the South American coast in what is now Venezuela.

As we have seen, around this time medicinal properties tended to be ascribed to virtually anything bitter. For the next fifty years and more, there were fruitless searches for cheap alternatives to the cinchona or Jesuit bark from Peru, using various other bitter plant extracts, such as coffee grounds. The new bark from South

America, like cinchona, was pleasantly astringent. Unlike cinchona, it does not contain any quinine, and has no pronounced curative properties; it does, however, have the cardinal virtue of not being actively harmful, and its extract is sold to this day for flavouring food and drink. It took its name from a region of Venezuela around the city now called Ciudad Bolivar, but which was then known as *Angostura*.

As with nux vomica, Angostura bark appeared on the European markets well in advance of any firm knowledge about its origins. Some thought that it came from a Magnolia. Eventually, Angostura bark was traced to a tree that had been given at least three names by different botanists: *Cusparia febrifuga*, *Bonplandia trifoliata*, and *Galipea officinalis*. This initial uncertainty about its origins did not prevent the market for Angostura bark growing strongly, so that by 1797, forty thousand pounds of the bark were exported to Europe in a single year.

In 1803, the state physician of Hamburg, Dr. Johann Jakob Rambach, was called in to investigate the death of a child who had been given a decoction of the new drug Angostura. He went to various apothecaries' shops, bought several samples, and found that the Angostura being sold in Hamburg was in fact a mixture of two different barks. The two kinds could be distinguished by careful examination or by chemical tests, but especially by their taste. True, or West Indian, Angostura was "aromatically bitter or astringent, succeeded in some degree by an aromatic flavour like mace." The impostor, on the other hand, was "in the highest degree disgustingly bitter, very durable, and not at all aromatic or astringent." It seemed to Rambach, although he had no direct evidence at first, that this second kind of bark was highly poisonous. As a result of his investigations, a decree was issued on May 11, 1804 informing all citizens of "this good town" of the dangers and threatening Hamburg apothecaries with a fine of fifty reichthalers if they sold any Angostura without official inspection and approval.

Sporadic fatalities from the "false Angostura" began to occur all over Central Europe, in Hungary, Latvia, Vienna and St. Petersburg; inspection of apothecaries warehouses in several countries showed that almost every Angostura sample was contaminated. Russia followed Hamburg's lead with an official proclamation telling citizens how to distinguish the true from the false. The Austrian court, after obtaining proof of the poisonous effect of false Angostura on animals, went further and ordered the destruction of all Angostura bark, both genuine and spurious, throughout the Empire, and forbade its importation.

The botanist Sir Joseph Banks ventured the opinion that the false Angostura did not come from the West Indies at all, but from a tree growing in Africa known as *Brucea antidysenterica* or *Brucea ferruginea*, named after the French explorer Jacques Bruce. Several years later, in 1817, it was still not common knowledge where false Angostura was coming from, or what it was—a situation that seems quite extraordinary to us today. An anonymous reviewer in the *Edinburgh Medical and Surgical Journal*[1] naïvely stated that although Dr. Rambach and others had accepted that false Angostura came from India, he was not so certain. Because both kinds often occurred in commerce mixed together, "it is more probable that both come from the same country, than that the importers and druggists should generally mix barks coming from such distant countries, however similar in appearance." But Edinburgh was a long way, both geographically and ethically, from the drug warehouses of London.

According to the German physician Husemann's highly xenophobic account,[2] the trail of the false Angostura led back first to Holland and then to England. The dim English, he said, imported a large quantity of bark from the East Indies, which John Bull had no idea what to do with. The problem was solved by passing it on to the "crafty and unscrupulous Mynheers" (an insulting term for the Dutch), who were responsible for the actual adulteration without any regard to the lives and safety of people in other lands; although Husemann had to grudgingly admit that no one had any idea at the time that the two kinds of bark had such dangerously different properties. The druggists, he said, did not have the decency to admit to the physicians that they were supplying false, or East Indian, Angostura, which they could buy in at half the price of the genuine article.

Not long after their isolation of strychnine, Pelletier and Caventou turned their attention to false Angostura. From this extremely bitter bark, they expected very probably to isolate strychnine, for it was by now known that false Angostura bark closely resembled nux vomica in its poisonous properties. However, after subjecting it to extraction procedures similar to those they had employed for nux vomica seeds and *Bois du Couleuvre,* Pelletier and Caventou obtained not strychnine but something resembling it but quite certainly different. The new substance was more soluble than strychnine and thus harder to purify. By now, however, they were gaining considerable expertise in the purification of their natural products, and found that the salt of the new alkaloid with oxalic acid is very sparingly soluble in water and alcohol, and were able to use this property to good effect. The insoluble oxalate salt could be decomposed by treating it with lime (calcium oxide) or magnesia (magnesium oxide) to give the new alkaloid. At this point they specifically stated, possibly for the first time, the techniques of crystallisation. If an alcoholic solution of their new substance was rapidly evaporated, the new alkaloid was produced in the form of mushroom-like lumps, but if the crystallisation was slow, and particularly if aided by the addition of a little water to the alcohol, they obtained perfectly regular crystals. After first considering the name *Angosturine* for the new alkaloid, they settled for *Brucine* because Angosturine might be thought to refer to the unrelated true Angostura, with the consequent danger of perpetuating the confusion. At this point, they thought that the bark was the bark of the African tree *Brucea.*

The new crystals had a powerfully bitter taste, but less intense than that of strychnine—persistent, sharp and acrid. Unlike strychnine, they melted without decomposition and at a much lower temperature. At this stage, the usefulness of an accurate melting point was not yet recognised; they did not make a precise measurement, although they did state that it is "slightly above that of boiling water." (Brucine melts at 105 degrees centigrade.)

Brucine was poisonous, but less so than strychnine, weight for weight, by a factor of about twelve. After carrying out a few experiments on dogs, the authors were unable to detect any difference from strychnine in its mode of action, except for this difference in potency.

To add to the confusion, it was not long before it was shown that the botanical identification of false Angostura had been erroneous. Already in 1808, as a result of investigations Rambach had carried out during the poisoning epidemic, he had suggested that false Angostura bark must come from nux vomica or a close relative, and in 1837 this was proved conclusively. False Angostura was none other than snakewood, the bark of either the nux vomica tree *Strychnos nux-vomica* or one of its close relatives such as *Strychnos colubrina*. Pelletier and Caventou had therefore isolated strychnine and brucine from different parts of the same plant. Both occur together in all parts of the tree, and the isolation of the substance they called strychnine on the first occasion, and of brucine on the latter, was partly due to higher concentrations of brucine in the bark and of strychnine in the seeds. The seeds contain about 0.5% or up to 1.5% of strychnine and perhaps 2% brucine, but because strychnine is more highly crystalline it is easier to isolate. Indeed, the bark of the nux vomica tree as well as the seeds had been used in Indian medicine, and it had reached the European market for many years past under the name of *Lignum colubrinum*, resembling the wood once described by Garcia da Orta. The identity of the various trees from which lignum colubrinum came was once again not completely certain, but most of it seemed to come from the nux vomica tree.

Lignum colubrinum or snakewood was often sold on the Calcutta markets as a substitute for yet another kind of bark, the innocuous rohun bark from *Soymida febrifuga*, a traditional Indian remedy for fevers. Another episode of mass poisoning which could have been even more damaging than the Continental outbreak was narrowly avoided by the accidental intervention of a Mr. Piddington.[3] About this man little is known except that he isolated what he thought was a new alkaloid, which he called "Rohuna," from a sample of what he thought was rohun bark. But the more talented pharmacist Dr. O'Shaughnessy[4] proved that the bark was nux vomica bark, or lignum colubrinum, and that the supposed new alkaloid rohuna was in fact nothing other than strychnine. This evidence was produced just in time to prevent a large consignment of the "rohun bark" being sent to the British Indian army for the treatment of fevers; one of several occasions on which strychnine knocked up against world events and changed, or might have changed but did not, the course of history.

Despite the misidentification, the name Brucine was retained for the second new alkaloid.[5] In the years to come, it would follow strychnine around like a spectre, a potentially complicating factor whenever it made its appearance in court.

Pelletier and Caventou's original sample of strychnine was not pure because subsequent investigators showed that the intense red colour with nitric or sulphuric acid that they described for their sample of strychnine is caused by brucine, not strychnine. Pure strychnine gives just a yellowish colour with nitric acid, and no colour at all with sulphuric acid. The attribution of colour reactions to strychnine which are actually characteristic of brucine was persistent because of the presence of brucine in most samples of commercial "strychnia." Thirty-seven years after the first isolation of brucine, during the Palmer trial, William Herepath was to describe the colours produced by treating strychnine with sulphuric acid and various oxidising agents, and to say that there was another class of tests used for "common strychnia"; that is, strychnia containing brucine.

Brucine became a byproduct from the manufacture of pure strychnine for medicinal purposes. It was cheap; it was also a product without uses, looking for a market. Most medicinal samples of strychnine contained a greater or lesser amount of brucine, and were thus less potent than pure strychnine. Brucine, or a mixture of brucine and strychnine, could be passed off as pure strychnine, with fatal results. The life of a patient treated with such an impure sample, whose doctor then unknowingly switched to pure strychnine, was forfeit.

Endnotes

1. *Edinburgh Medical and Surgical J.*, 1817; 13, 209.
2. Husemann, 1857.
3. Pereira, 1855: Chevers, 1870:242 refers to "The late Mr. Piddington, Coroner of Calcutta," and this was presumably the same person.
4. Sir William Brooke O'Shaughnessy (1808–1889), who was professor at the Calcutta Medical College and incidentally the first person ever to attempt laying an underwater telegraph cable, under two miles of the Hooghly river in 1839.
5. The name Caminarine is also recorded but soon disappeared from use.

CHAPTER 6

You Will Be Careful as to the Second Article

The contradictory deductions of medical professors make wise men tremble, good men sad, and bad men bold...

A.S. Taylor, *Poisoning by Strychnine*, 1856

In 1848 Great Britain was nominally at peace, but it was a turbulent year. There was revolution in Berlin and at home the Chartist riots continued. The Irish famine was in full spate. Five thousand miles of railway lines had been built, a third of them already equipped with the electric telegraph. Powerful social changes were at work in the English shires; Britain was undergoing the massive changes that would make it the first country in the world to become more than half urban. Soap was taxed and most people did not wash very much. In rural southern counties such as Hampshire, the standard of living had not improved in a generation. Many families lived on a diet of bread, cheese and beer, and were dependent on the local parish rates to supplement their subsistence wages.

All of this may not have unduly worried Mrs. Sergison Smith of Romsey in that county, who was comfortable enough. According to the *Hampshire Independent*,[1] she was "an amiable and beautiful lady in the prime of life" (she was about 35). An officer's wife and mother of five, she lived in an agreeable house, called Jermyn's, and employed several servants, one of whom, Caroline Hickson, was described as a "nurse and lady's maid."

On October 30, 1848, Mrs. Smith, who had recently suffered a miscarriage, felt unwell with "weakness" and the local pharmacist, Mr. Jones, sent in a prescription of nine grains of salicine (willow bark), a useful analgesic, mixed with orange-peel. This arrived at about six o'clock in the evening. The hypothesis that Mrs. Smith was not really ill is supported by the fact that she did not take any of it until seven o'clock the next morning, when Hickson saw her drink half a wineglassful. She said that it seemed to taste unlike previous prescriptions of the same medicine that she had been given by other druggists, but she thought that this may have been because she had also been taking laudanum to treat a pain in her face. The laudanum had given her bad dreams during the night, and she had been terrified by a vision of a madwoman.

Five or ten minutes later, she called for Hickson who found her still in bed but leaning on a chair at the bedside. Hickson thought she had fainted, and she appeared to be suffering from spasms. Hickson sent the coachman to fetch the surgeon, Mr. Francis Taylor, and when she got back she found Mrs. Smith lying on the bedroom floor surrounded by the other servants. She was screaming, very much, very loudly, Hickson said. Her legs were drawn up and contracted, and she begged to have them pulled straight, and to have water thrown over her. Hickson noted that her feet were turned inwards. She put a hot-water bottle on them, but it had no effect. The governess who was also present said that she had seen her mother affected in the same way, and Mrs. Smith said through clenched teeth, "Oh, you have, have you; do you think it is hysteria then?"

A little while later, Mrs. Smith said that she felt easier, and asked to be turned over. A few minutes later she died. She was conscious throughout, and recognised Hickson to the last. From the time she took the medicine to her death was about an hour and a quarter. Her teeth were clenched the whole time. At a later hearing Hickson was asked, "Was she the whole of the time from the fit coming on stiff?" "Yes," she replied, "she was only relieved a very few minutes before she died."

Taylor testified that he saw Mr. Jones, the pharmacist, come running up the drive in a state of great excitement. Jones's wife told the newspaper reporter that he "Did not feel so much the consequences to himself as the thought of having sent such a beautiful creature to another world, and such a good customer." Shortly afterwards a "very respectable" jury met at the house to view he body and consider the evidence. It heard that the salicine had been kept in a small bottle on a high shelf of the pharmacy, on which there was also a bottle of something else; strychnine hydrochloride. Jones had dispensed the prescription himself and had muddled the bottles. He was arraigned on a charge of manslaughter and committed suicide not long afterwards.

After death, Mrs. Smith's hands were tightly clenched and her feet twisted; her eyelids adherent to the eyeballs. The unnatural stiffness dissipated slowly after death, but after three days the hands and feet were still distorted.

Even social parasites deserve better than this.

Throughout the nineteenth century and beyond, the ready availability of strychnine-containing medicines was a constant danger, and the medical profession was by no means a certain bulwark against disaster. The case of poor Mrs. Sergison Smith was not unique. In the following year, 1849, *Punch,* in the form of a spoof letter from Paracelsus Pillcock, M.D., laid into doctors and pharmacists for illegible and incomprehensible prescriptions.[2] Not long afterward, a young lady in London paid the price. She had been taking a nux vomica medicine for some time and in May 1853 her doctor wrote her a repeat prescription.

It read:

Pulv. Strychnos

Nucis vomicae ℈ij

Bismuth Nit. jss

Pulv. Rhei. gr.viij

The young woman's father was pressed for time and took the prescription to an unfamiliar pharmacy. He unnecessarily told the assistant, "You will be careful as to the second article, the nux vomica, and let it be good." The inexperienced assistant assumed from this that the prescription was for both strychnine and nux vomica and put in 1⅔ grains (Ⴢij) of each. The prescription proved fatal in one hour and a half. A verdict of accidental death with censure of both the doctor and the druggist was returned. *The Lancet*[3] opined that no single person was responsible for the mix-up. The word *Strychnos* was irrelevant; the term *Pulv. Strychnos* should not have been written on a separate line, and (as any Islington druggist's assistant should have known, the editors imply), *Strychnos* was not in the correct case (the genitive) to agree with *Pulvis nucis vomicae*. The assistant should have known that 1⅔ grains of pure strychnine would be fatal and should have queried it. But they could not refrain from bemoaning the fact that if only the self-important father had refrained from putting his oar in, the mix-up might not have occurred: "If lay persons would not pretend to more technical knowledge than they really possess, nor presume to read prescriptions and give directions guided only by a slender acquaintance with such writings...." It did not occur to them to speculate that if doctors refrained from writing their prescriptions in mediaeval cabalistics, neither the father nor the assistant would have been confused.

The extent of nux vomica's use in traditional Indian medicine is uncertain. There are contradictory accounts. According to one nineteenth century source, it was seldom if ever used by Hindu physicians because of the dangers.[4] However, if a modern compilation of traditional remedies is to be believed, it would probably be easier to list those afflictions for which it has *not* been recommended.[5] Typically, in India the nuts were boiled in milk to soften them, then scraped into a paste which can be set aside and softened again when needed. They were also recommended as prophylactics against various diseases. It was commonplace to take nux vomica every day to protect against snake bites. In a paper read by Mr. Baker to the Medical and Physical Society of Calcutta in 1823, he related the case of a servant who, while carrying an umbrella for a British officer, suddenly fell down in a fit. He appeared stiff and lifeless, and when he stood up, rested on his heels, with his toes turned upwards, his eyes staring, and his teeth clenched so strongly that the company could not separate them. After a few minutes he recovered, vomited, and told Baker that he had been taking the Kuchila, or nux vomica, for four months to guard against rabies, starting with an eighth of a nut and gradually increasing the dose so that he was now on two a day, taken morning and evening. That morning he had taken it on an empty stomach.[6] Baker reported that the nut was taken "either coarsely powdered, in its natural state, or half roasted on a hot iron, changing it from side to side till it swells out, which it does, in the course of one or two minutes." Two nuts would certainly contain a fatal dose of strychnine, although the roasting would decompose some of it. Later research would claim that although strychnine is not a cumulative poison, neither is it an habituating one; continually ingesting it leads to little or no tolerance, so the manservant was daily dicing with death.

Following Fouquier's work during the opening years of the nineteenth century, nux vomica, and very soon afterwards strychnine, showed a meteoric rise in popularity not only in France but throughout Europe, and before long in North America too. The use of nux vomica, together with that of other plant drugs, was described by Magendie in his influential *Formulaire Pour la Préparation et L'Emploi de Plusieurs Nouveaux Medicaments* (1821), and within a very short space of time the efficacy of nux vomica, and very shortly thereafter of strychnine, had become received opinion among French physicians.

By the later 1820s, a succession of snippets in *The Lancet* and elsewhere were reporting Continental physicians' claimed successes. At about this time, Pelletier and Caventou founded a laboratory in Paris for the commercial preparation of alkaloids, and their example was followed in London by Thomas Morson. In 1828 there appeared the first English-language recipe for the preparation of tinctures and pills of nux vomica and pure strychnine. This was based on Magendie's formulary, which had already gone through five editions in French, each translated into English.

In no time at all, a succession of British doctors had jumped on board the bandwagon. Deafness, headache, intestinal worms, prolapsed rectum, lead poisoning (or *paralysies saturnines* in French), rheumatism, diabetes, catatonia, strangulated hernia, even cholera; all these citadels and more were subjected to the cannonade of the new drug. M. Bernard at l'Hôpital de la Pitié in Paris was operating on partially blind patients by injecting strychnine sulphate powder into the eyelid with "a small pen-shaped penetrating instrument."[7] In a treatment that makes the eyes water even to think of it, urinary retention was treated by injecting a solution of strychnine up the urethra into the bladder using a probe.

Another influential publication popularizing the new drugs was the long-lived *Treatise on Therapeutics* by the wonderfully named duo of Trousseau and Pidoux.[8] Armand Trousseau began his career as a teacher of rhetoric until he was persuaded to take up medicine by the physician Bretonneau. The first edition of his book with Pidoux was published in 1839 and the last, the ninth, containing additional sections on magnetism, electricity, acupuncture, massage, gymnastics and flagellation, almost forty years later. Throughout this period, the section describing the use of nux vomica and strychnine was virtually unaltered. Reading between the lines of the text makes it clear that the drugs' reputations had been founded from an early date on competing and dubious claims by various French physicians seeking to outflank Fouquier and gain credit for the claimed therapeutic successes which always persisted in being never quite clear-cut. Trousseau and Pidoux in their first edition say:

A more or less complete knowledge of the physiological action of this 'Héroique substance' led Fouquier to recommend it in paralysis; and although he himself was less successful, perhaps, than others have been, we must still award him the honour of the discovery....Initially he used it against hemiplegia, but Bretonneau repeated his experiments and soon found that while it had little value in hemiplegia, it may be given with great profit in paraplegia, and in general in paralyses which depend on diseases of the cord, or of the nervous conductors only. After many trials [Bretonneau] formulated the cases in which it ought to be tried as follows; paraplegia symptomatic of concussion of the medulla, when the primary symptoms are past and paralysis alone remains; those which follow inflammation of the medulla or its membranes, when all the phenomena of local irritation have long disappeared; those which follow Pott's disease; the various paralyses caused by lead...[9]

Poor Fouquier! He introduces a nostrum which he thinks will work, gets equivocal results which he describes honestly, then has the credit taken away by less fastidious colleagues, such as Bretonneau (the man who got Trousseau his job, and to whom therefore he owes a favourable mention) who says that Fouquier would have succeeded if only he had used it against the right *kind* of paralysis.[10] Bearing in mind the power of autosuggestion on patients given such a violent treatment and the fact that many neurological diseases are self-limiting or get better spontaneously, it was easy for any doctor to find a set of patients to support his account of successful cures. It is unlikely that any of them misreported their results mendaciously; it was easy enough to be persuaded unconsciously by the climate of general consensus that such a violent drug must do *something* useful.

The case could not have been put more clearly than by the pen of Trousseau himself. He is actually referring to the lack of hard evidence concerning strychnine's claimed efficacy against cholera, but in fact what he says gives the game away in respect of the entire strychnine saga:

> ...from these results, important as they perhaps are (unfortunately this is still very doubtful), how great is the distance to...that kind of infallibility which was at first attributed to the strychnia-treatment, an illusory infallibility, which soon disappeared before failures both numerous and marked.

By 1833, a minute but telling change in nomenclature had already surfaced. Dr. R. Rowland of Fenchurch Street, London, was using "The Strychnine" to treat neuralgia, period pains, amenorrhoea and hysteria in women. Clearly strychnine had passed from being an interesting, if chancy, new remedy, to becoming a minor institution. Dr. Rowland had never seen any ill effect from its employment, nor did he believe that it required "such watching as is usual to inculcate." By 1835, at least two books had already been published in England on the miraculous properties of the new drug,[11] and in 1836 it was included in the new London Pharmacopoeia, apparently against the wishes of some doctors.

Some physicians were administering stupendous amounts of nux vomica powder in a regimen of gradually increasing doses which would leave the life of the patient hanging on the question of how efficient his digestive system proved in absorbing the poison, how fine the powder was, and how much strychnine it contained. The situation was made even more hazardous by the widespread adulteration of drugs. Second-rate nux vomica powder adulterated with substances such as ground date-stones was often supplied to doctors, with lethal results. In France, strychnine was adulterated with chalk, magnesia, sugar or amidone: a criminal offence.[12]

Trousseau, now using strychnine rather than nux vomica (but in what degree of purity?) was recommending hair-raising doses. After giving the recipe for a syrup containing approximately 2 milligrams of strychnine per 5 millilitre teaspoonful, he describes how to administer it:

> We give on the first day two or three teaspoonfuls of the syrup, according to the age, insisting on its being given at equal intervals during the day, at morning, noon and night, so that the effect may be watched and the aim not overshot. If the dose of three spoonfuls is well borne, it is continued at first for two days, then increased by one spoonful; then wait two days more, and thus attain the amount of six, always giving each at the proper intervals.

> When this dose is reached, we substitute a dessert-spoonful (representing twice as much as the former) for one teaspoonful, and following the same rules, gradually reach six dessert-spoonfuls per diem....

The mind boggles. Six dessertspoonsful, or 24 milligrams per day, is nearly half the fatal dose, generally reckoned as one grain or 62 milligrams, and as little as half a grain taken all at once may have killed a 39-year-old man. But the real horror is yet to come:

> ...[thus] we at last give children of from five to ten years of age from 25 milligrams to 6 centigrammes [60 mg.] (0.4– 1.0 grains)...

So these are children's doses!

> Beyond this age we begin with the dessert-spoonful, and gradually reach 10 centigrams (1½ grains) of the active principle, for adolescents...It is the more essential, to watch over

the treatment, as the remedy must be given in doses large enough to betray itself by physiological action. It is necessary, also, to let the person in charge know what may happen...

In a very few days, when we begin to increase the original dose, the patient feels at certain moments in the day, twenty or thirty minutes after the dose, a little stiffness of the jaw, headache, disturbance of vision, a little vertigo....In some cases, there are also muscular shocks and often (if there is hysteria) spasms or convulsions. These shocks are produced especially when the patient is surprised, as when he receives a sudden order, and may throw him to the ground. The contractions are tetanic, painful, especially when there is an effort to resist and to remain upon the feet, but lying on the flat of the bed is sufficient to quiet the storm at once.

What better way for a doctor to retain complete control over a fractious patient; better than an electric fence, and all in the name of science too!

What is a poison? In his influential book *A Treatise On Poisons* (1829)[13] Robert Christison, who had studied in Paris at the time of the isolation of strychnine, declined to give a definition and pitched straight into describing the various poisons and their effects. He contented himself with dividing them into two main classes: the irritant poisons, which destroy the tissues at the site of administration, and what he called "the nervines," poisons that act at a remote site in the body.

Had Christison wanted to come up with a good working definition, he could have done worse than go back to Paracelsus, who in 1538 wrote:

What is not a poison? All things are poisons, and nothing is without toxicity. Only the dose allows anything not to be poisonous. For example, every food and every drink is a poison if consumed in more than the usual amount, which proves the point. I admit that a poison is a poison; but that is no reason for condemning it outright.

In other words, "It's all a matter of the dose."[14] Foods, drugs and poisons are all foreign chemicals taken into the body with varying results, however much this may offend believers in natural foods and remedies. Alcohol is a good example of a substance that clearly fulfills all three functions, depending on the dose; strychnine is certainly both a drug and a poison, although it has to be admitted that any role it plays as a food is hardly in evidence.

From the earliest days there had been accidental poisonings by nux vomica. Matthiolus in about 1550 told of an old lady killed by eating cheese made in a bowl in which it had been bruised. Later on, some drunken apothecaries in Germany bet one of their numbers that he would not be prepared to drink beer containing *Cocculus indicus*[15] which they said would make him walk on his head. The man accepted the bet, but for a joke presumably, was given nux vomica instead. He went home and died fifteen minutes later.

Paradoxically, nux vomica powder, although less potent than pure strychnine, was the more dangerous. By the time they reached Europe, the extremely hard nuts could not be ground by hand, only filed. Physical factors—the fineness of division of the powder and its exposure to stomach acid—would strongly affect the degree of absorption of the strychnine, and were insufficiently taken into account by the early physicians. Fine powdering would later become possible with steam power and milling machinery.[16] The spread of steam boiler technology in the first half of the nineteenth century allowed not only mechanical milling, but also the use of steam itself to soften the nuts so that they could be more easily powdered.

The problem was that when someone was given nux vomica powder, a high proportion of the alkaloids was retained in the nut tissue after it was swallowed, and leached out into the body only slowly. It was always possible to find unchanged powder in the stomach, from which a large amount of unchanged strychnine could be extracted.[17] Thus, although it is usually stated that strychnine is not a cumulative poison, nux vomica certainly could be. One lady was given three grains of the powder three times a day for sixteen days. She began to suffer from purging and colic, so the treatment was stopped. Five days later, she complained of ringing in the ears, drowsiness, increased sensitivity to light and sound, numbness and deafness. On the ninth day after stopping the treatment, she became speechless and convulsions set in. The symptoms increased and she died, "apparently exhausted," four weeks after first being prescribed it, and a full twelve days after ceasing to take it. Nux vomica powder was therefore an extremely dangerous cumulative poison with wildly unpredictable effects; the most dangerous medicine that has ever been administered to human patients, despite some stiff competition.

From an early date, the more sensible doctors, including Fouquier, preferred an alcoholic extract of nux vomica, which was in theory at least more reliable and controllable than the powder, and then, as it became widely available, strychnine itself. But we have already seen how commercial samples of the alkaloid were very often contaminated with the less active brucine, and were also subject to adulteration, so the doctors had only very rough-and-ready ways of controlling the administration of such a perilous substance. Assaying, that is determining the effective strength of, such crude plant extracts so that the dose could be given accurately was a major problem to nineteenth-century physicians; or at least, to those that thought about it at all.

There are contradictory accounts concerning whether pure strychnine can act as a cumulative poison. This property is of signal importance when considering the medical evidence in the famous poisoning trials, such as the Castaing and Palmer cases. Whilst medical authorities are unanimous in saying that strychnine is not cumulative, its behavior in these cases was more unpredictable than their opinion would imply. The explanation is probably that when good scientific studies of strychnine metabolism were carried out and its fate in the body properly probed later in the century, the alkaloid was administered by injection or intravenously. But given by mouth, there is a complicating factor which I have not seen specifically referred to by any of its early enthusiasts. Strychnine is a nerve poison which interferes with muscle control and although these effects are most noticeable in the skeletal muscles of the limbs and chest, the nerves supplying the internal organs could also have been affected.[18] This would have affected the functioning of the digestive tract, and is probably partly responsible for the intermittent nature of strychnine poisoning. Small doses were said to stimulate the intestines, but larger doses would have thrown them into spasm, thus interfering with the further absorption of the poison. The patient would partially recover as his digestive tract lost its power, then as it recovered a fresh surge of strychnine would be absorbed and the spasms would start again. So although strychnine given by injection is not a cumulative poison, strychnine given by mouth probably is. Whatever the reason, strychnine administered orally was an unpredictable tiger caught by the tail.

Trousseau and Pidoux, despite their blind attachment to administering the drug, did at least recognize this property:

> When these effects appear we must not increase the dose any further, for strychnine, like all the preparations of nux vomica, is a remedy which, by virtue of its very peculiar long therapeutic range and a most remarkable accumulative action, is capable of causing unforeseen accidents, even though it has been given in moderate doses which up to a given point of time produced scarcely appreciable effects...tolerance of the remedy not only varies according to individuals but even in the same person, so that while the doses remain the same, we never can infer tomorrow's effect from that of yesterday; violent spasms may occur tomorrow immediately after the first dose, even when we are sure that the preparation remains the same.[19]

If given by mouth, a drug is potentially liable to chemical change as soon as it is swallowed, since it is subjected to the strong stomach acid, followed by the alkaline environment of the intestine. Assuming it is not affected by these and is absorbed into the bloodstream unchanged (or is given by injection), it may be metabolised by the tissues, especially the liver, before or after it has done its work. Some drugs (prodrugs) are not in the active form when they are administered; it is the metabolised form that is effective.

Some very powerful drugs, such as anticancer drugs, react chemically with their target, such as the nucleic acids of the cell. Penicillin reacts chemically with the cell walls of bacteria. But most drugs carry out their action without chemical change. They fit a receptor somewhere in the cell (the famous "lock-and-key" mechanism), cause their physiological change without themselves being altered, and are then released. This question of whether a drug can take effect without being chemically altered in doing so became of vital importance in the Palmer trial, as we shall see

Strychnine is chemically stable and it is the unchanged alkaloid that is the active drug. It is however gradually metabolised by the liver, with a half-life of about 50 minutes; that is, half of a given dose disappears in each fifty-minute period after it reaches the bloodstream. So when it kills, it does so within an hour or two. The fatal dose of pure strychnine is generally given as ½ to 1 grain, or approximately 30 to 60 milligrams, but as little as 15 milligrams has killed. For comparison, a typical aspirin tablet contains 300 milligrams of active ingredient. So, strychnine is a highly dangerous poison and was especially hazardous in the nineteenth century when given in doses such as those described by Trousseau. His doses are so enormous that the strychnine with which he made up his syrup could not have been pure; even without conscious adulteration, it must have been a weaker (and variable!) mixture of strychnine and brucine.

Even from the first, there were sceptics. An 1822 review of Magendie's book[20] says that it was easy to see how such an energetic substance could soon become a favourite treatment, and in Paris, its use had become as indiscriminate as Magendie could possibly have wished. But the reviewer had good reason to know that many of the patients relapsed as soon as they staggered out of the hospital gates. "Surely," he says, "when we possess so many pleasant, efficacious and innocent bitters... it is an effort of the blindest attachment to propose the substitution of so formidable and disagreeable a remedy." Thirteen years later, the anonymous reviewer of Mart's book[21] preferred sarcasm:

> Thirty-two more miracles performed by strychnine! Mr. Mart has come from H.M. Ship *Raccoon* and has discharged a very destructive broadside against hemiplegia, paraplegia, amaurosis, nervous indigestion, tic douloureux and neuralgia...the present work is not—Oh, fellow countrymen! a list of the *killed* and *wounded*. It resembles more a report of the Humane Society, and contains only a catalogue of *the cured*....We ask Mr. Mart, and we shall continue to ask all the proclaimers of new medicines—"To how many patients

did you administer your drugs without producing the least advantage from its employment?"

Mr. Hawkins of St. George's Hospital had not much faith in strychnine and believed it to be useful only in cases of temporary paralysis, and had never seen it to produce any better effect than spasmodic twitching of the limbs. He does not quite go so far as to query why it was necessary to use it at all, if it only affected temporary paralysis, but that is the implication. A Dr. Granville tried it out on an intimate friend of his, an eminent chemist, for a long period, but the treatment completely failed.[22] Dr. Epps, using it in 1836, had noticed unfavourable effects such as convulsive movements of the legs, and as a result had transferred his patients to a much safer regime: arsenical liquor.

Others were more motivated by fears about its safety than doubts about its usefulness. Sigmond[23] bemoaned the fact that there was so much variation in the strengths of nux vomica preparations and said that if this could be overcome, there would be no necessity to use strychnine, "which is the most ferocious poison we possess, and which, I think, should never be prescribed." Despite endorsements by Magendie and other famous physicians, it should be, in his opinion, "shunned by the practitioner." Others agreed, following a meeting of the Medico-botanical Society, that Strychnine's use was so dangerous that it should be confined to the most urgent and hopeless cases.

The history of strychnine, as it developed, reflected the different social structures of the two dominant European nations. France was (and remains) a highly centralized country, often with one set of received opinions about scientific questions. After the researches of Magendie, Fouquier and the others, this opinion was solidly behind strychnine and it is impossible to find a dissenting voice. Medical opinion in Britain was more difficult to corral. There were always those prepared to take a contrary view, and although there are no signs of overt anti-French xenophobia in the negative comments about strychnine, the fact that the new ideas came from over the Channel made it especially easy to be sardonic. But once a certain critical mass of medical opinion had fallen into line, the unrestricted capitalism of Victorian Britain came into play and the positions were reversed. When the dangers as well as the supposed benefits of strychnine were realised, the French climate of regulatory control ensured that its use was restricted to the medical elite, while in Britain, through the offices of the numerous unregulated drug wholesalers and apothecaries, before long anyone with twopence to spend could lay his hands on it.

There are early signs of hysterical overtones. Sigmond informs us that, "a curate, who had been in the habit of using strychnine," (he does not say what for), "presented a scarcely appreciable morsel to a denuded surface, and was instantly killed."[24] The morass of confusion surrounding drugs, electricity, nervous phenomena and mesmerism persisted. Did not the limbs of a corpse twitch when electricity was passed through, and did not strychnine produce the same effects? It was widely thought that anyone touching a strychnine-poisoned human being or animal received an electric shock. A Mr. Moore from Mitcham wanted to destroy a dog and locked it up in a room with a saucer full of strychnine dissolved in milk. When he returned to the room some time later, the dog was dead *despite not having drunk any of it.*

But the bandwagon rolled on. Later claimed uses tell us much about Victorian obsessions. For urinary incontinence in children of both sexes, strychnine provided a useful alternative to cold baths, blistering and cupping of the loins. J.H. Houghton in 1856 is getting to grips with a really serious ailment. "In the daily round of practice there are few cases more annoying to the medical attendant, or less satisfactory in their general management, than habitual constipation," he writes. Medicines such as colocynth and ipecacuanha work at first, but gradually require greater and greater doses. A lady, age 55, corpulent, inactive and self-indulgent, had not gone a single night for eight or nine years without taking nux vomica, "except when she has been ill, and required other treatment." Another successful case had been of a patient whose life had been "dominated by colocynth, calomel and an enormous bottle of senna mixture and had become one of great misery, for the little comfort she got whilst her bowels were allowed to remain quiescent was destroyed by her fears of the next necessary resort to medicine." He does not report what form of words he used to persuade her not to be frightened of nux vomica.

A few years earlier, M. Duclos was treating another bogus but anxiety-causing nineteenth-century condition, spermator-rhoea.[25] Nineteenth-century doctors found difficulty in distinguishing between sexual potency and fertility. Overuse of the sex organs, either deliberately or involuntarily, could result in loss of spermatozoa and thus infertility. On the other hand, complete abstinence in men, and especially women, had all sorts of undesirable side-effects including infertility and most notoriously hysteria, defined literally as disorder of the womb.

In the treatment of spermatorrhoea, patients were given nux vomica pills in gradually increasing numbers. The treatment could be supplemented with nux vomica ointment rubbed into the loins and the inside of the thighs. Duclos writes that the condition of incomplete impotence is as often found in those who have been excessively continent as in those who have abused their sexual organs. Self-pollution may occur either by night or day, the discharge being either a true or a pseudo-spermatorrhoea, whatever that is. Commenting on Duclos's work, the editor of *The Lancet* reveals that he has many times tried strychnine in these lamentable cases, but has never discovered any special benefit, and prefers cauterisation as the treatment.

As the century progressed, some formulation of either nux vomica or strychnine became a virtual obsession with some physicians. Strychnine pellets became so handy for one American doctor that "I can give them to a businessman to carry in his vest pocket to take one or two occasionally for indigestion or headache caused by the sluggish action of the liver, stomach or bowels....Railroad men, farmers etc. can carry them and use them in hot weather when they are inclined to drink so much water that it hurts them...children can take a few pellets to school in cases of bowel trouble when they are not sick enough to stay out of school..." Victorian physicians can remind one of a motorist stuck far from a garage with a car that won't start. Their range of techniques is strictly limited. "Why don't you make sure the plug leads are all on properly?" a passer-by says. They try this, and occasionally it works. If this happens, they make sure never to set out on a journey without first hammering on the leads so tightly that sometimes they break. For some physicians, strychnine became the WD-40 of Victorian medicine.

In 1856 a cogent plea for the deletion of strychnine from the pharmacopoea appeared.[26] J. Nichol FRCS pointed out that many highly poisonous medicines, arrested or guided by the hand of science were useful, but this was not the case with strychnine. In his opinion, not one of the effects that it had exhibited in its rapid career justified its use. "That it may, in very minute doses, in combination with other agents, produce some beneficial effect...may be possible, although I am inclined to doubt it; but these effects are equally obtainable from other medicines, and certainly do not afford an excuse for its use," he goes on to say. "The frightful disclosures being made in the courts and the careless manner in which this deadly poison passed from hand to hand make it a subject worthy of the attention of the medical profession and of the legislature."

Dr. Nichol, had he survived long enough, would have been dismayed by the number of accidental fatalities from strychnine poisoning and by the illicit uses to which it would continue to be put over the next few decades, often by doctors themselves. No doubt he would have been gratified to know that his wishes would eventually be fulfilled, in only a little over a century.

We will let a Scots physician have the last word. Sir James Mackenzie (1853–1925), in his time a well-known Harley Street cardiologist, tells us that he resolved to try its action for himself:

> Strychnine had a great vogue, its effects being commented on by physicians and surgeons—the latter, indeed, often refusing to operate on a patient under an anaesthetic unless the patient had a preliminary dose of strychnine, while anaesthetists had often beside them a hypodermic syringe ready charged, lest the heart would fail. To my surprise I could get no result with strychnine whatever....no effect could be found on the healthy heart, nor on people in a state of collapse. I read up the literature, and beyond assertions as to its value, there was not on record a single instance where there was given any evidence justifying the belief in its properties.[27]

This assessment, made in the early years of the twentieth century, probably puts Mackenzie among the earliest of those whose change of opinion contributed to the painfully slow decline in strychnine's medical reputation. But before that took place, a lot of nux vomica would flow under the bridge.

When all else had failed and the patient was on the way out, strychnine could be used to resolve things, one way or the other. "At one time few patients in Britain were allowed to die without being given strychnine injections."[28] Not surprisingly, the biography sections of the nation's libraries can be scoured in vain for any specific mention of strychnine being used as this treatment of last resort.[29] The physician was restrained by professional ethics; the friends and relations would not have known what the last injection contained, and as for the patient himself...

How many famous Victorians left this world with the words ringing in their ears, "there is just one other thing that I might try..."?

Endnotes

1. *Hampshire Independent*, Nov. 4, 1848:5–6. The spelling of her name is as given in this newspaper report. Other spellings such as "Serjeantson Smyth" appear elsewhere.
2. Punch, 1849:37. "Pillcock" may have been a spoof on Dr. Locock, who achieved some fame as a marketer of Dr. Locock's Powders for All Disorders of Children (Stuart, 2004: 84). A Dr. Locock treated Helen Abercrombie when she was dying in 1830 (see Chapter 9).
3. *Lancet*, May 17, 1856:550.
4. Chevers, 1870:243.
5. Nadkarni, 1954:1175 et seq.
6. Sigmond, 1837.
7. This and subsequent similar descriptions of strychnine treatments are taken from the *Lancet*, circa 1830–1835.
8. Trousseau and Pidoux, 1839; Trousseau and Pidoux, 1875–1877.
9. Ibid, 231.
10. Fouquier sounds like an engagingly modest man. "I have been so much credited with this discovery that I only made by chance," his obituarist records him as having said (Requin, 1852).
11. Bardsley, 1830; Mart, 1835.
12. Chevallier, 1850.
13. Christison, 1829.
14. Albert, 1987.
15. Indian cockle or fishberry is another Indian plant. It is a narcotic and used to stupefy fish so that they can be caught. It contains picrotoxin (not an alkaloid; it does not contain nitrogen).
16. As late as 1803, the Apothecaries Company installed horse-driven mills for grinding medicines at Apothecaries Hall in London. Strong heating partially decomposes strychnine and brucine, and reduces

their concentration in nux vomica to a less dangerous level. In traditional Chinese medicine, the seeds are heated in dry sand or in very hot soya oil, then powdered (Boo-Chang Cai, 1990).

17. Taylor, 1859.
18. Although glycine receptors are commonest in the central nervous system, they also appear to exist in the autonomic nerves controlling the digestive tract.
19. Trousseau and Pidoux, 1876–1877 (transl. Lincoln), vol. II.
20. *Edinburgh Medical and Surgical Journal*, 1822, 18, p.15.
21. *Lancet*, 1835:112.
22. This was almost certainly Michael Faraday, a lifelong hypochondriac. Dr. Augustus Bozzi Granville (1783–1872) was one of the proposers for Faraday's membership of the Royal Society, and was a fellow-member of the newly founded Athenaeum club (James, 1991, vol. 1, 315 (footnote), 341).
23. *Lancet*, 1837:826.
24. Ibid.
25. Involuntary discharge of semen. A condition publicised by Claude François Lallemand (1790–1853), professor at Montpellier, who advocated circumcision as the cure. His ideas were taken up enthusiastically in Britain and the United States. One quack showed a patient with the aid of a microscope that his urine contained spermatozoa, and told him that he was condemned to death unless he handed over £50 for medicines and spent the next 28 weeks locked up in the dark (Acton, 1870). In the nineteenth century many cases of presumed "spermatorrhoea" would have been discharges caused by gonorrhoea. Believers in medical progress should, however, see how many Internet sites they can count today prescribing homeopathic cures for spermatorrhoea.
26. *Lancet*, Sept. 1856:29.
27. Wilson, 1926.
28. *Pharmaceutical J.*, 1962:152. "A wise physician would know when to ease the administration of powerful stimulants like digitalis and strychnine" (Jalland, 1996; 92, citing a 1917 doctor).
29. The only explicit mention I have so far come across is in an account of a remarkable (and unsolved) twentieth century poisoning case. In 1928–1929, three members of the Sidney family were separately poisoned over a period of nearly a year with arsenic at an address in Croydon. The doctor found one of them in a state of collapse and gave him injections of digitalis and strychnine, without effect. Connoisseurs of the poisoner's art will be gratified by the additional fact that the third of the victims, Violet Sidney, was killed by someone putting arsenic into her bottle of Metatone, the strychnine-containing "tonic" (Symons, 1960: 163–171).

CHAPTER 7

You Hold Him Down, I'll Pour It Down His Throat

Everyone now knows that when a man is accused of having poisoned another...the poison must be detected, and it must be proved that the accused has used this poison to accomplish his criminal intention. In times in which the means of detecting poisons with the greatest certainty were not yet known, the rack was used to make the discovery...

Justus von Liebig (transl. W. Gregory), 1851

Although deliberate poisoning was well known to the Arabs and in Renaissance Italy, it was always rare in England, although there were mentions. In the fourteenth century, the chronicler Ranulf Higden claimed that Eleanor of Aquitaine, wife of Henry II, had poisoned Henry's mistress Rosamund de Clifford a century and a half earlier in 1176.[1] But there is no firm evidence for this. Poisoning in the Middle Ages was often by repute, when a chronicler wished to assign particularly base motives to someone.

The pace hotted up during the turmoil of the Tudor period. In 1532, the cook to the Bishop of Rochester, Richard Roose, added "a certain venym or pyson" to the porridge and caused the illness of seventeen of the guests and the death of one of them, Bennett Curwen, gentleman. In addition, the village people waiting at the gates for the leftovers were also poisoned, and "one pore Woman, that is to say Alice Trppyn, wydowe is also thereof now deceased." It has been suggested that the poisoning was an assassination attempt on the bishop, John Fisher, by friends of Anne Boleyn, because he had opposed the divorce of Henry VIII from Katherine of Aragon. Whatever the motive, Parliament was so outraged that it passed a law prescribing execution by boiling alive for all poisoners. This sentence was carried out on Roose and one or two others before it was repealed in 1547. Hysteria about the possibility of poisonings persisted for several decades. In 1594 the Portuguese Jew Roderigo Lopez, who had settled in London and became the first physician at Saint Bartholomew's Hospital, was hanged at Tyburn for plotting to poison Queen Elizabeth.[2] In the same year, Edward Squyer was executed by disembowelment after placing his hand on her saddle and crying out, "God Save the Queen," having obviously placed there a Spanish poison powerful enough to kill at the merest touch.[3]

Poisoning was always far more widespread on the continent than in England. By 1659 it came to the attention of Pope Alexander VII that waves of Italian women were confessing that they had poisoned their husbands. The mania throughout all strata of society for disposing of unwanted relations and rivals spread from Italy to France, and in the latter country led to strict controls on the possession of poisons. Although there is some possibility that famous Italian poisoners, such as the witch Hieronyma Spara, may have used nux vomica among other things,[4] it is impossible to be certain about this, and in any case, as we have seen, what was often described as nux vomica was more likely to have been the thorn apple, which could be grown locally.

In India there was a well-documented tradition of nux vomica poisoning. The bands of thuggees wandered the highways, and later the railway system, tricking travellers into taking a poisoned meal so that they could be robbed. Their most common drug was the thorn apple, or dhatoora, which rendered the victims insensible for several days and frequently killed them or destroyed their mental powers. But Hervey in 1861 reported to the Foreign Office that the kuchila or nux vomica as well as arsenic and other poisons were also used. In 1888 there were 360 recorded cases of thuggee poisoning in Bombay alone; "And no wonder," Dr. Mair, the coroner for Madras reported, "Probably in no other part of the world, did greater facility exist for procuring poison either in the bazaars or from gardens—arsenic, corrosive sublimate, opium, stramonium (dhátura), and other deadly poisons, are openly sold in the bazaars in any quantity, the sellers being unrestrained by any license." As the author, a bemused colonial administrator, comments, "In no country in the world...has the progress of civilization been [so] disfigured by the contemporaneous existence of bands of plunderers by hereditary descent...in India, we have to deal with criminal systems which have been [*sic*; seen?] the growth of ages, and with criminal deeds the depths of which are utterly inscrutable."[5]

Several factors may have contributed to the lack of any well-documented record of nux vomica as a deliberate poison in Europe before the nineteenth century. The most likely is that until the 1830s, arsenic was so attractive to the poisoner that there was hardly any need to look further. Another factor was that nux vomica was not widely available. And a third may have been that its public image had always been that of a handy poison for animals, and its possible use as a deliberate human poison may just not have occurred to people. Although its highly toxic nature was well known by the time of John Gerard in the seventeenth century, it continued to be thought of as a dog poison. Giovanne Porta described it as such in 1558.[6]

The Venoms that Kill Dogs

Nux vomica which from the effect is called Dogs Nut, if it be filed, and the thin filings thereof be given with Butter or some fat thing to a Dog to swallow, it will kill him in three hours space; he will be astonished, and fall suddenly, and dies without any noise; but it must be fresh, that Nature seems to have produced this Nut alone to kill dogs.

This description of "Dogs Nuts," taken together with other passages in his book, make it easier to understand why nux vomica was not seen as a human poison. There was very little appreciation in Porta's time of what we today would call taxonomy, the view that some species might be more closely related than others. To him, each species had its own God-given spirit and constitution that set it apart from other plants and creatures. The fact that a goat might mate with a sheep, or a horse with an ass, is no more nor less remarkable than that a tiger might mate with a crocodile. With our modern knowledge, we would tend to assume that if nux vomica poisons dogs, it would very probably (though not necessarily) also poison cats or humans, although we would not be particularly surprised if it did not poison, say, insects. To Porta, every species was distinct, with its own mythology. Nux vomica was something put on the earth with which to poison dogs, just as in the same way, crushing basil leaves caused scorpions to appear, but not spiders or serpents (for which you needed to crush the hairs of a menstruating woman).

> Do not think I mean, that one poison can kill all living Creatures, but everyone hath his several poison; for what is venome to one, may serve to preserve another; which comes not by reason of the quality, but of the distinct nature.

Even when books began to be written about forensic investigation, nux vomica was still referred to primarily as a dog poison. According to an 1820 book, there was even some doubt about whether it was poisonous to humans,[7] although this may have been deliberate disinformation.

Another explanation for the lack of any real historical record in Europe concerning deliberate poisoning by nux vomica is of course that from the seventeenth century onwards it was used, but either the miscreants got away with it, or the authorities, mindful of the dangers, colluded in wiping all mention of it from the record.

Criminologists agree that poisoning is a fairly rare form of homicide, accounting for perhaps 2% of cases. Since it is normally a domestic crime, once poisoning has been proven as the cause of death, the finger of suspicion invariably points to the person in the family with the motive; it is rare that there are multiple suspects, whatever Agatha Christie might say. Hence the clear-up rate for poisonings is high.

Or is it? The poisoner's main hope is to conceal the fact that poisoning has taken place at all by passing it off as death by natural causes. One lawyer in the mid-nineteenth century held the opinion that for every poisoner convicted, six escaped detection, but how he arrived at this figure is difficult to guess. The reformer William Farr wrote in 1839, "It is not improbable that a certain number of cases of poisoning escape undetected by coroners and juries who can be expected to know little of the symptoms either of poisons or disease...the prospect that the effects of poisoning may be confounded with natural causes, offers a strong temptation to the commission of that dreadful crime."

An important contributing factor to this unsatisfactory state of affairs in England was the professional position of the coroner. Although the post was established as far back as 1194, by 1800 it had become one of ridicule, with incompetent coroners carrying out inquests in village inns. They were quite incapable of tackling the legal implications of serious criminality, much to the disgust of radical thinkers like Dickens, who described such a travesty of an inquest in *Bleak House*. Coroners were not expected to be medically or legally qualified. But a greater defect was the ludicrous attitude taken towards suspicious, as opposed to violent, deaths. From as long ago as 1487, coroners were paid 13/4d ($1.50) for each inquest on a slain body, but not if there was no visible cause of death. Three hundred and fifty years later the situation was, if anything, worse. The better coroners began to realize that they needed help in the form of expert witnesses, but the regulations stipulated that they had to be paid out of the Coroner's own pocket. He then had to argue to reclaim the money at the next quarter-sessions, often without success. The case of *R. vs. Kent* (1809) centred on such an attempt by Kent county magistrates to disallow inquest fees in a case where, in their opinion, there had been "no ground to suppose that the deceased had died any other than a natural, though sudden, death." Chief Justice Ellenborough endorsed their action. He said that it constituted a salutary check against the "many instances of coroners having exercised their office in the most vexatious and oppressive manner, by obtruding themselves into private families to their great annoyance and discomfort, without any pretence of the deceased having died otherwise than a natural death, which is highly illegal."[8]

When one Julius Pampe died suddenly in 1839, the radical MP and progressive coroner Thomas Wakley, a medical man, gently tried to direct the jury in the direction of an autopsy. While saying that "There could scarcely be any doubt as to the cause of death, since Pampe came from a respectable family and there were no suspicious circumstances;" nevertheless, he told them, "If you have any [doubt] it can be removed by causing a post-mortem to take place." But the jury, while "warmly commending the views of the coroner, but having no doubt in their minds," returned a verdict of natural death.[9]

A worse case occurred a few years previously. A young woman from somewhere near London, crossed in love, went out one evening to the shops. On returning to the house, she begged her mistress for forgiveness, then died of dreadful convulsions. A local druggist had sold a young woman two drachms of nux vomica earlier that evening, but was unable to positively identify her. At the inquest, the parish surgeon said that he had opened the stomach and extracted a brown powder. This had been tested by a medical gentleman, who was in the room and would testify as to its composition. Mindful of the rules of evidence, he did not go on to say what the doctor had found, but waited for the coroner to call for this evidence. He waited in vain. The doctor was not even called, and the coroner directed the jury to find a verdict of "Died by the visitation of God." Any attempt to give you an account of the examination of the medical evidence by the non-medical coroner would beggar all description," The Lancet's correspondent reported, "Was ever anything more monstrous!"[10]

Lord Campbell, who was Lord Chief Justice in the 1850s and presided at the trial of William Palmer, did not have a very high opinion of coroners. He said publicly that evidence given at the inquests on some of Palmer's victims could not be relied upon in court because coroners' inquests were often defective. As The Lancet pointed out, this was an unfortunate remark given that Campbell himself was the chief coroner for England, and had only just finished getting legislation through Parliament to improve the competence of coroners! But in the real world, Campbell's remarks were sound. It would have been difficult to devise a system more poisoner-friendly. The murder of newborn infants was widespread, and was encouraged by the existence of burial clubs, primitive life-insurance schemes which paid out the undertaking costs. The epidemic of child-poisoning reached its zenith in 1846–1851 when the abuses led to the closure of the societies. "Sometimes a matter of a pound or two, nay, of a few shillings only, have been enough to cause the wild sacrifice of a human life...with little more compunction than a grazier would exhibit in disposing of his flocks for the shambles." Most of these poisonings were carried out in a "grossly bungling" way with poison bought from the local shop. One perpetrator asked the shopkeeper for some arsenic "to kill rats, and not to poison any body with."[11] But many adults too must have been surreptitiously poisoned. Sudden death from infectious diseases such as cholera was common. It is true that arsenic is a far better mimic of cholera than is strychnine, but the odd death from "apoplexy," or "epilepsy" would not have gone amiss from time to time; it was not until the Palmer trial that the symptoms of strychnine poisoning were thoroughly aired in a public forum. As Taylor pointed out, the only difference between burial clubs and life insurance was the affluence of the clientele.

Britain thus lagged well behind Continental Europe in recognising the need for a separate discipline of forensic science, or Legal Medicine or Medical Jurisprudence as it was initially known.

The very idea of forensic toxicology, the investigation of a death for proof of the poison used, was first contemplated in 1781, when J.J. Plenck said, "The only certain sign of poisoning is the botanical character of a vegetable poison or of the chemical identification of a mineral poison found in the body." As far as the practical application of this precept goes, Plenck was ahead of his time; 45 years ahead in the case of arsenic, and about 70 years in the case of strychnine. Mathieu Orfila, who was born on Minorca in 1797, is generally considered the founder of modern forensic toxicology, although by the year of Orfila's birth, Fodéré had already published a large textbook on the subject in France, which by 1813 had expanded to six volumes.[12]

In 1821 the young Robert Christison witnessed Orfila lecturing to 1,000 students at the École de Médecine in Paris. But although the first British chair in the subject was established in Edinburgh in 1807, nothing further happened until 1830, when the Society of Apothecaries made it part of its licence requirements. This prolonged reluctance was a reaction against Continental ideas of state interference, as conservative factions in England saw it. Soon after Christison returned from Paris, he was appointed to the Edinburgh chair, wrote his excellent book on poisons, and became the leading British expert.

For many years, arsenic (strictly, arsenic oxide) was the poison of choice for surreptitious poisoners. It is colourless and tasteless, so there was never any difficulty in administering large amounts of it. But by the 1830s, Orfila and others had developed reliable tests for detecting minute amounts of arsenic.

As his or her preferred poison became readily detectable, the arsenic enthusiast had a stark choice. He (or she, for a disproportionate number of poisoners have always been women) could persist with arsenic, in so many ways the ideal poison, and conceal the fact that there had been any poisoning at all, so that the authorities would not test for it. Or he could turn to something else; the vegetable poisons, such as strychnine, morphine or nicotine, which continued to be untraceable. The discovery of pure strychnine could not have come at a more opportune time for the poisoner; shortly before other advances in science made arsenic detectable. Here was the essence of nux vomica in handy crystalline form, widely available (at first, admittedly, only to physicians, but later on to all and sundry), and it was undetectable. True, it was difficult to administer because of its extreme bitterness, and it produced symptoms unlike those of any natural disease except the rather rare tetanus.

As late as 1847, Orfila thought that vegetable poisons might remain forever undetectable.[13] But only four years later, a Belgian medic turned chemist, Jean Servais Stas, successfully isolated and identified nicotine in the stomach of a poison victim, Gustave Fougnies.[14]

Fougnies was poisoned at the Chateau de Bitremont near Bury in Wallonia, on the French–Belgian border. His assassins were Count Hippolyte de Bocarmé and his sister, who had been expecting him to die soon and leave them his money, but had discovered with consternation that he had decided to get married. The Count, a scion of the local nobility whose father had been governor of Java, was an enthusiastic amateur botanist. He grew tobacco plants in his greenhouse and distilled the leaves, telling his simple-minded gardener that he was manufacturing eau-de-cologne. On the day in question, the Count held Fougnies down on the dining-room floor while his sister, who claimed she was acting under duress, poured the distillate down his throat, followed by some vinegar to disguise the smell. This was a mistake. Stas realised that Fougnies had been poisoned when he saw burn-marks round the mouth caused by the combination of nicotine and vinegar; vinegar alone would not have burned the skin.

He therefore successfully developed the method known as Stas distillation for isolating the alkaloid from the stomach contents.

The Stas procedure, which worked only for liquid (steam-volatile) alkaloids, was improved by Julius Otto in 1856. After the initial digestion of the stomach contents with alcohol and acid, Otto used ether to extract out fatty substances. When the acid was then neutralised, the alkaloids, no longer present in their water-soluble salt form, passed into an ether layer which could be washed with water to remove impurities and evaporated to give much purer samples of the offending alkaloid than previously available. On evaporating the ether onto a white porcelain tile, there was obtained a yellowish ring of impure alkaloid surrounding a central zone containing pure white crystals. If strychnine was suspected, the colour test could be applied by adding a drop of sulphuric acid, then a single crystal of potassium dichromate to give characteristic rings of blue colour. This colour test, and variations of it, remained the standard forensic method for detecting strychnine for the rest of the century.[15] The preparation of the sample for the colour test and the carrying out of the actual test were operations of great delicacy requiring an experienced experimenter. The more skilled investigators became adept, but the lingering doubt always persisted as to whether a procedure based on a chemical reaction that was not understood (because the chemistry of strychnine itself was not understood) could be completely reliable. The picture was complicated by successive interventions of people who lacked the necessary experience in carrying out this extremely delicate test, failed to get a result, and erroneously claimed that the test was unreliable, or not as sensitive as the experts claimed.

Despite its naissance in France, strychnine almost disappeared from use there. Inorganics seem always to have remained the poisons of choice for French poisoners; as arsenic became easier to detect, phosphorus (from matches) became the leading French poison. Dupré and Charpentier[16] ascribe this to the fact that "only those people well instructed know the use of these alkaloids," and few cases involving strychnine occur until much later in the century. The *Annales d'Hygiène Publique et de Médecine Légale*, which recorded all cases of poisoning in France, did not contain a single case of strychnine use in the 27-year period from its establishment in 1829 until 1855. "This difference has certainly an explanation in the legislation which, in England, gives liberty to the sale of poisons whereas, with us, the restricted sale of poisonous substances only allows a small number of poisons with industrial or economic use, among which strychnine is not included."[17] Much later, in 1895, we find M. Roy of Troyes (a pharmacist) poisoning his former mistress Alice Adamsky with strychnine-containing chocolates sent through the post with the help of his current friend Rose Estrat.

W.H.G. Remer's book on forensic chemistry, first published in Prussia in1812,[18] describes laurel, opium, scammony resin, jalap, gum-gutta, sabine, ciguë, belladonna, black henbane, the mysterious Aqua Tofana (probably an arsenical decoction), and lead acetate, but there is no mention of nux vomica. Fodéré gives a much longer list.[19] But again, nux vomica is conspicuous by its absence. Neither does it appear in a pleasant little book published in 1823 in England about household poisoning.[20]

A clue is found in Fodéré's book. He makes it clear that conscious censorship, as well as tight legal control, was now being imposed in France. Certain recent scientific advances in the field are not to be communicated to the general public. This is just before the publication of Pelletier and Caventou's isolation of pure strychnine, but after Magendie's studies of the powerful toxic effects of nux vomica, by which it is known that an immensely powerful poison is lurking in the beans.

But the concept of strict regulation to control such dangerous substances was slow in catching on in England. By mid-century the only poisonous substance controlled by law was arsenic, the sale of which was regulated by the highly unsatisfactory Arsenic Act of 1851, which did not prevent the fatal accidental poisoning of 20 people at Bradford in 1858. No legislation would be passed controlling other dangerous poisons such as strychnine until the Pharmacy Act of 1868, and even this proved capable of circumvention by the ingenious Christiana Edmunds.

Between 1852 and 1856 there were an average 13,711 violent deaths each year in England and Wales. Of these, 401 were poisonings. The majority of poisonings were accidents or suicides, with opium the most frequent cause of death. The most popular agent for murder was arsenic. There were 83 trials between 1739 and 1878 at the Old Bailey for poisoning,[21] of which 49 were for homicide and 31 for attempted homicide. (Trial for manslaughter by poison is impossible, since it is a premeditated crime.) Of these, three cases involved the use of strychnine. Later, between the abolition of public hanging in 1868 and the end of the century, there were 337 hangings for murder throughout England and Wales, of which 22 were for poisoning.[22] Of these, nine involved strychnine. In the slightly shorter period from 1900 to 1929 the number of executions increased to 500, but only nine of these were for poisoning, and only one was with strychnine. So the cases described throughout this book represent the majority of *known* homicidal uses of this poison in England.[23]

But were there other, previously unrecorded cases? Even when Thomas Cream was tried in 1892, the procedures for regulating sales of such a horrendously poisonous substance as strychnine were still very lax. John George Kirby, a chemist of 22 Parliament Street, was cross-examined about sales of nux vomica when Cream came into his shop several times between early October and December 20, 1891, buying between one and four ounces each time. As this was a scheduled poison, Cream was asked for his name and address and for a written order on each occasion. Kirby could only find two of the purchase records, one just dated "20th," and the other undated; they did not need to be entered in a register; "it comes under the second schedule and so long as it is labelled poison with the name and address of the seller it is quite sufficient. For a sale across the counter we would not book it." Kirby asked Cream if he were a medical man, and looked him up in the Medical Register, but did not find his name there. When asked if he was in the habit of selling poisons to persons he could not find in the register, he replied no. He could not explain why he had not registered the purchase when he could not find Cream in the register, and could not say whether other chemists followed the same practice.

Almost the same concerns had been raised earlier at the trial of Dr. Lamson for poisoning by aconitine, which he had bought at Messrs. Allen and Hanbury in November 1881. An assistant at the pharmacy told the judge that if an ordinary member of the public wanted to buy a scheduled poison, "they would have to be introduced by someone known to us." If, however, they were satisfied that the customer was a doctor, they just referred to the Medical Directory to see if there were such a person. The judge asked him, "Supposing I went in, and having got hold of the Medical Directory and taken a name-say Mr. Brown—would you serve me with two grains of aconitia?" "That would not be sufficient," the chemist replied, "You would have to write your name in a formal manner."

"Then do I understand that anyone of respectable appearance and well dressed might apply, and that without any means of satisfying yourself that he is not an impostor and not telling you what is untrue you would supply him?" "The only test would be the style of writing which is characteristic of medical men." The procedures for registering and monitoring doctors were for many years similarly lax. At the Cream trial, the secretary of St. Thomas's said that no Dr. Cream, or Dr. Neill, was connected with the hospital "so far as I know." Although it was "usual" for any gentlemen attending the hospital to report to him, "it sometimes occurs that old students may be attending without registering their names with me."[24] Regulations on the sale of poisons were tightened up following the realisation that doctors were just as likely to be poisoners as any group of people; indeed, more so, as a small percentage of them were likely to have been attracted to medicine because of its morbid opportunities.

Furthermore, they could get their hands on the right stuff.

Endnotes

1. Weir, 1999.
2. Berger, 1995–7:79.
3. Wilson, 1989; Thompson 1925.
4. Stuart, 2004:118. The author is in error in ascribing the use of nux vomica to the famous poisoner the Marquise de Brinvilliers. As far as I am aware, current opinion is that she used lead acetate, known as the "Poudre de succession de Brinvilliers." Poisonings describing vomiting and/or a slow death with stomach cramps can be discounted as nux vomica cases.
5. Hervey, 1892.
6. Porta, 1558. But see also note on page 244 that dogs cannot taste strychnine.
7. Orfila and Black, 1820.
8. Burney, 2000.
9. Forbes, 1985: 14.

10. *Lancet*, 1833-4, [1] :97.

11. Interview with Alfred Swaine Taylor, *Illustrated Times*, 2 February 1856. But see also Mrs. Major in Chapter 17.

12. Fodéré, 1813.

13. In the trial of Wooler at the 1855 Durham assizes, the defence unsuccessfully argued that because the accused had a packet of strychnine in his pocket he could not possibly have poisoned the deceased with arsenic, the poison found in the body, because everyone now knew that arsenic could be detected. Therefore if he had wanted to commit murder he would have used strychnine...

14. Bouchardon, 1925; Vandenbussche and Braeckman, 1976.

15. *Liebig's Annalen*, 1856: 100, p. 39. This refinement came just too late for the Palmer and Dove trials. Otto mentions in his paper that he has read the press reports. An alternative, or corroborative back-up, to the colour reaction for strychnine was to prove that the extract would kill a small animal. In India in 1870, killing one of the abundant lizards was recommended as an easily carried out test for the alkaloid (Chevers, 1870:245).

16. Dupré and Charpentier, 1909.

17. Tardieu, 1857: 60.

18. Remer, 1816.

19. Foderé, 1813; "Nous terminerons ici cette section des poisons en particulier, qui est déja trop long, si l'on ne considère que le danger qu'il y a faire participer le publique de certaines connaissances qui ne devrait être qu'entre les mains d'un petit nombre de sages.... j'ai même omis de parler d'un grand nombre se substances veneuses et métalliques nouvellement découvertes." These sentiments were echoed more than forty years later by Alfred Swaine Taylor; "I could add largely to the list of poisons which either by their nature or by their tremendous power in very small doses....might infallibly produce death without leaving a physical or chemical trace of their presence in the body. I forbear to do this." Christison was less guarded. Toward the end of his life (he died in 1882) he prepared two popular lectures on poisons. There was clearly concern about this in Edinburgh, and he defended his action by saying that although poisons could be used to murder, it would be valuable for the general public to know what to do in the much more numerous cases of accidental poisoning. Discretion seems to have won the day, for the lectures were never delivered (Christison, 1885). Perhaps as well. According to an anecdote (Thompson, 1925:71), Christison was once giving expert evidence in a trial, in the course of which he said to the judge "My lord, there is but one deadly agent of this kind that we cannot satisfactorily trace in the human body after death, and that is-." The judge interrupted him; "Stop, stop, please, Dr. Christison. It is much better that the public should not know it." Christison would refer to the incident in his lectures at Edinburgh, and tell the students that the undetectable poison that

he had wanted to refer to was, in fact, aconitine. An assiduous note-taker in the class one year was George Lamson, who was hanged a few years later for poisoning his brother-in-law with aconitine.

20. Anon, 1823c.
21. Forbes, 1985
22. Fielding, 1994–5.
23. A Scottish case is fully described by Roughead (Roughead, 1938: 69–101).
24. Shore, 1923.

CHAPTER 8

Overture to the Sorcerer's Apprentice

SCENE: *The shop of Mr. Upas, Chymist and Druggist. BOT-TLES, his Assistant, is behind the counter...*

BOTTLES: *Now, Sir, What can I do for you?* [*To a Stranger, with his face muffled and his hat over his eyes*]

STRANGER: *Thank'ee: I'll wait.*

BOTTLES [*to several customers*]: *You for Arsenic—You?—You?—All of you Arsenic? Six Arsenics; and you? Oh! One corrosive sublimate.* [*Exeunt with the poison*]

STRANGER [*having watched them all out*]: *I want some of the strongest poison you have got.*

BOTTLES: *Well, Sir, I think Prussic Acid will suit you better than any.*

STRANGER: *That smells, doesn't it?*

BOTTLES: *Why, yes, sir. Probably strychnine would answer your purpose?*

STRANGER: *Is that pretty stiffish?*

BOTTLES [*smiling*]: *Why, yes, Sir. I should be very sorry to take two grains of it.*

STRANGER: *Let's have half an ounce.*

BOTTLES: *Half an ounce, Sir?* [*weighs it out*] *What is the next article, Sir?*

STRANGER: Nothing.

BOTTLES: Allow me to tempt you with a little Belladonna; very killing Sir, I assure you. Or would you try our digitalis? I could recommend the Colchicine, Sir.

STRANGER: Nothing.

BOTTLES: Or anything in the vitriol way, Sir?

STRANGER [with an oath]: No, I tell you. The Strychnine will do the job. Hand it over, will you, and make haste.

BOTTLES: Directly sir. Thank you, sir. [exit Stranger] Ha! A pretty good morning's work, and if the undertakers don't get a job or two out of it—and perhaps Jack Ketch too—I shall be astonished rayther. [Exit]

Punch, July 1849

There are at least two interesting things about this playlet. Firstly, its description of how easy it was to buy strychnine and other poisons in the Victorian pharmacy is hardly an exaggeration, and it is a telling comment on the slapdash state of affairs that enabled the famous poisoner William Palmer to flourish. But the second, and much more remarkable thing, is this: had it been written and published during the public hysteria and indignation of the Palmer furore, it would have been merely a piece of Mr. Punch's heavy-handed wit in rather poor taste. But, extraordinarily, it was written and published in 1849, seven years *before* the Palmer trial. If Palmer was the first Englishman to stand trial for murder by strychnine, what was going on in the years between 1820 and 1856? It is quite clear that the *Punch* writer knew that strychnine was readily available, was notorious for its poisoning powers, and had almost certainly been used to murder people. But, more to the point, *the readership knew it too*. There is no point in publishing a lampoon that means nothing to the audience.

In France, poisoning had been widespread in the seventeenth century, whence it had spread from the Italy of the Borgias. The authorities clamped down. An edict of 1682 prohibited the sale of poisons to non-householders.[1] The seller was required to keep a careful register, each page countersigned by the Lieutenant-General of Police. Poisons were kept under the key of the proprietor, who alone was allowed to dispense them. From 1777, only qualified druggists were allowed to sell medicines, and they were not allowed to sell anything else. The authorities visited each establishment at least once a year and infringements were met with heavy fines. Another law prohibited medical men from dispensing their own prescriptions.

Paradoxically, it was these stringent controls as a result of the former widespread lawlessness that made France on the whole a safer refuge from poisoning during the eighteenth and nineteenth centuries than England, where there had never been any controls at all and complacency won through. Even after the Palmer revelations of 1856, the *Pharmaceutical Journal* felt able to say, "Such precautions fortunately have not been found necessary in this country."[2]

The first known cases of deliberate strychnine poisoning took place in England; a shadowy, doubtful affair in the early 1830s and the well-documented *cause célèbre* of Palmer in 1856. Paris, though, was the birthplace of strychnine. It was available there before it could be obtained anywhere else, but only by medical men.

In this chapter we will look at a celebrated poisoning in France from the beginning of the period in question that might just have involved the alkaloid, and in the next chapter we will turn to the notorious affair in England a few years later that probably did.

In 1823, Dr. Edmé-Samuel Castaing,[3] age 27, residing at no. 31 Rue d'Enfer, Paris, was arraigned in court for the murder of two brothers and for the destruction of the legal will of one of them.

Castaing was born at Alençon in Normandy in 1796, the youngest of three sons of a forestry inspector.[4] After schooling in Angers, this intelligent and hard-working middle-class youth decided to become a medical student, and moved to Paris. Here, about two years into the course, he fell violently in love with the widow of a judge, who already had a family.[5] Soon after he qualified as a doctor in 1821, they produced another two children, and it has always been assumed that his subsequent crimes were the result of his genuine love for her, a determination to provide for her, and the need to pay off the debts that accumulated.

Castaing was friends with the brothers Claude-Louis-Auguste and Daniel-Hippolyte Ballet. The elder of the two, Auguste, was a man of dissolute habits who had become estranged from his family; legend had it that when a baby, he had been let fall by his nurse, and although he was not seriously injured, from that day on his mother had developed an irrational coldness towards him. Hippolyte, the younger, had been his mother's favourite but was a consumptive in delicate health. Both parents were now dead and the two brothers had both inherited money. There was also an elder half-sister, now Mme. Martignon.

In 1822 Hippolyte Ballet, suffering from hypochondria and depression, consulted various doctors who advised him to take the waters at Enghien for his tuberculosis, advice which he followed. In September of that year, he returned to Paris, where he was treated by Castaing. On October 2nd Mme. Martignon visited him and found him well, with a good appetite, but during the night he suffered severe attacks of vomiting. The next day his brother-in-law M. Martignon found him in bed with a swollen face and red eyes; the same evening Mme. Martignon was turned away by the servants, who told her that Hippolyte was resting and that Dr. Castaing had been with him all day. During the next two days she made several attempts to see him, but was refused by Castaing, who claimed that the sight of her would be too upsetting for the patient, even when at one point she tried to get into the room wearing a servant's clothes so that Hippolyte would not recognise her. After further suffering, Castaing always present and denying access to visitors, Hippolyte died on the morning of Sunday the 6th. A post-mortem was carried out by Castaing and another doctor, and his death attributed to pleurisy and consumption.

An extraordinary amount of financial dissimulation now enters the picture with the tangled history of the will that Hippolyte may or may not have made, and the fate of the money that he left when he died. The previously poverty-stricken Castaing suddenly became the possessor of 100,000 francs, apparently as a result of Auguste writing to his stockbroker on the day of his brother's death requesting him to liquidate 100,000 francs of his assets, send the money in cash and destroy the letter. What apparently happened was that Castaing got hold of Hippolyte's will in which he had left most of his money to his sister Mme. Martignon. Castaing told Auguste that the will was in the possession of M. Lebret, the deceased father's former clerk and an informal legal advisor to the family. He told him that M. Lebret (who seems to have been totally honest) had demanded 80,000 francs from Mme. Martignon to hand the will over, but that for 100,000 francs Castaing could outbid her and get Lebret to destroy it, in which case Hippolyte would be deemed to have died intestate and Auguste would inherit all the money. Auguste withdrew the 100,000 francs, the two drew up outside Lebret's office in a carriage, Auguste Ballet gave Castaing the money, he went inside and came out a few minutes later showing the seals of the will that he claimed he and Lebret had destroyed, but which in fact he had already destroyed earlier.

The ruse was a brilliant one. Castaing now had complete power over Auguste Ballet, who thought that they were co-conspirators in the destruction of the will. Since he thought that Lebret was corrupt and that the 100,000 francs had gone to Lebret not Castaing, Castaing could claim that he had not benefited from the crime and could now threaten Auguste with exposure.

Before long, Castaing was asking the advice of his cousin Malassis, a notary's clerk, as to whether a will made out by a sick man in favour of his doctor would be valid. Soon afterwards, Malassis received the will of Auguste Ballet naming Dr. Castaing as his sole beneficiary, with a covering note from Castaing asking him to look after it.

On the evening of May 29, 1829, after a visit together to the country, the two returned to Paris and took a twin-bedded room at the Tête Noir hotel in Saint Cloud. The following day they went out walking and returned to the hotel to dine at seven. Then they went out again and went up to their room again at about nine, ordering some warm wine to be sent up to them. Castaing mixed it with some lemon and sugar that he had obtained. Auguste complained about its bitter[6] taste and got his servant to try it also.

Later Auguste suffered from pains and swelling, and in the morning could not get his boots on. During the night, Castaing had got up and disappeared for two hours, between four and six a.m. During this period he arrived at the door of a Paris apothecary and asked for twelve grains of tartar emetic, which he said was "For Dr. Castaing." At seven he visited another chemist, from whom he had bought morphine acetate during Hippolyte's illness, and bought 36 grains of it. He returned to the hotel later in the morning, asked for some milk for Auguste, then left again, after which Auguste suffered violent pains and vomiting. Castaing wanted to send to Paris, a distance of ten miles, for a doctor, but Auguste insisted on someone local, and consequently a Dr. Pigache was sent for. He visited three times during the day and prescribed soothing medicines; Castaing told him that he thought Auguste was suffering from cholera; when he asked to see the vomited material, he was told that it had been thrown away on Castaing's orders. Pigache proposed returning later that night, but Castaing said that it would not be necessary, and they agreed on a further visit the following morning.

At about this time, Castaing sent a note to Jean, Auguste's black servant back in Paris, telling him to deliver the keys of Auguste's desk to Malassis. But Jean knew about the will in Castaing's favour and distrusted him; accordingly, he wisely went to Saint Cloud to deliver the keys personally to Castaing.

When he arrived, Jean found his master very ill. Late that night, Castaing gave Auguste in Jean's presence a dose of the medicine prescribed by Pigache. Four or five minutes later, Auguste suffered violent convulsions. Dr. Pigache was sent for, together with Professor Pelletan from Paris. They found him unconscious with his eyes open, his neck stiff and his pulse weak, racked with periodic spasms. Pigache bled him and applied twenty leeches. He died the following midday. Castaing impressed all with his display of crocodile tears; "I am losing a friend of my childhood," he said, and both the parish priest and the clerk who accompanied him went away impressed by his piety. The doctors, though, had found the death *"Extraordinaire et contre l'ordre naturel des choses."* They pointed out the awkwardness of the situation, for Castaing had told them how distressing it was for him to have been present at the deathbed of both brothers, especially in view of his being their sole heir. "You are right," Castaing told Pelletan, "My position is dreadful. In my great grief I had never thought of it until now, but now you make me see it clearly. Do you think there will be an investigation? I beg you to insist on a post-mortem. You will be acting as a second father to me in doing so." One of the witnesses at Castaing's subsequent trial described him as looking more like a priest than anything else.

Suspicion, then, was immediate. Castaign made several errors pointing to his guilt; for example, in sending a note to Malassis instructing him to tell no one that they were related, and by at first denying that Auguste had written a will, then admitting that Malassis had it. He also showed immense interest in the post-mortem, hanging around the door hoping for news of the findings. When they emerged they were favourable to him; death, the doctors said, was due to inflammation of the stomach, which could have been caused by sunstroke or some other kind of over-indulgence; no trace of poison could be detected.

Nevertheless, Castaing was arrested and tried. During his imprisonment, he at first feigned insanity, then tried to communicate with the outside world through other prisoners to influence various possible witnesses.

The trial opened on November 10, 1823. In accordance with the French system, there was first read an immensely long arraignment containing all of the evidence against the prisoner together with every conceivable fragment of circumstantial evidence. At last the prisoner was brought in to be examined. The prosecution soon dropped its accusations concerning Hippolyte Ballet, for there was no evidence firm enough to secure a conviction. Eventually, Castaing was convicted on the charges of destroying the will and of poisoning Auguste Ballet (by a seven-to-five majority), and was guillotined.

All of the evidence was circumstantial, for there was no firm forensic proof. As would happen in the Palmer trial more than thirty years later, Castaing's team tried to defend the proposition that in the absence of forensic evidence no conviction was justified. But the *avocat-general* dealt with their submission caustically:

If the actual traces of poison are the material proof of murder by poison, then a new paragraph must be added to the Criminal Code—"Since, however, vegetable poisons leave no trace, poisoning by such means may be permitted with impunity." To poisoners I will say in future, "Bunglers that you are, don't use arsenic or any mineral poison; they leave traces; you will be found out. Use vegetable poisons; poison your fathers, poison your mothers, poison all your families, and their inheritance will be yours; fear nothing; you will go unpunished!" ....We have gone through a large number of facts. Of these there is not one that does not go directly to the proof of poisoning, and that can only be explained on the supposition of poisoning; whereas if the theory of the defence be admitted, all these facts, from the first to the last, become meaningless and absurd. They can only be refuted by arguments or explanations that are childish and ridiculous.[7]

The evidence is tenuous as to exactly how Castaing killed the Ballet brothers, but there is some circumstantial evidence that he used strychnine alongside other poisons. We need to consider the characteristics of each poison in turn.

Tartar emetic (antimony potassium tartrate) had been known since the Middle Ages and was considered important enough to have its own alchemical symbol. Introduced as an emetic by Adrien de Mynsicht at the beginning of the seventeenth century, it was at first used clandestinely by physicians, but gained respectability when it was given to Louis XIV as a child and was claimed to have effected a cure. It is a white powder, almost tasteless. Today it is considered far too hazardous to use in medicine. It is poisonous, but the danger lies more in its unpredictability than its potency. The fatal dose is relatively large and if administered alone, it fairly reliably induces vomiting.[8] It was taken medicinally on a vast scale, for example in James's powder for colds. These were invented by Dr. Robert James (1705–1776), who often prescribed his powders to be followed by a mercurial pill then a dose of "bark," which may often have been *Strychnos* bark. This could be a lethal combination, and it killed the playwright Oliver Goldsmith in 1774.

Morphine is bitter, but nowhere near as bitter as strychnine. It is a soporific and painkiller, fatal in large doses and killing its victims by sending them to sleep. Vomiting is listed as a frequent side effect. Convulsions can occasionally be produced at very high dosage, but this is unusual.

Strychnine, as we have seen, is intensely bitter and extremely poisonous, producing powerful convulsions. It is *not* an emetic.

At the time of the Castaing case, French physicians were in the midst of an intense debate as to whether emetics were or were not useful. In this as in other areas of medicine, they were handicapped by lack of true scientific knowledge, particularly of infection. Diseases such as cholera were marked by disorder of the gastrointestinal tract, but they thought that this disorder was the cause of the disease, not its result. Vomiting is a useful mechanism that the body has for getting rid of stomach contents that it does not like; but, according to the doctors who were against the use of emetics, their use exacerbated the disease by increasing the disorder of the stomach. To add to the complexity, they only had the one reasonably reliable emetic, tartar emetic, (often just referred to as "emetic"), which was itself toxic.

According to the court proceedings, Castaing was a friend of the physician Chevalier, whose 1820 thesis had dealt with just this point.[9] Emetics, Chevalier says, rather than destroying the illness, exacerbate the existing inflammation and are dangerous. If that is the case, as Castaing and Chevalier probably discussed, emetics can cause fatal disease if their effects are bottled up and allowed to continue for long enough. How to do this? Give something that calms the stomach and allows the disease to develop. The obvious candidate is morphine.[10]

The reasoning is faulty, but the effects are the same. By suppressing the vomiting reflex, morphine could allow tartar emetic to kill by poisoning; not because it is an emetic, but because it is toxic.

Where does strychnine come in? There is no hard evidence for Castaing having purchased any, or having had any in his possession. But if there was anywhere on the planet that it was available, it had to be Paris; Magendie's book recommending its use to physicians had already reached its third edition. The method of poisoning using tartar emetic and morphine was inefficient because Castaing's theory about killing by repressed vomiting was incorrect, and tartar emetic is only a moderate poison. So if Castaing had initially embarked upon this rather inefficient regimen for poisoning Hippolyte Ballet, it must have been extremely tempting for him to beef up the performance of the tartar emetic by introducing another, stronger, poison.

This is not pure speculation. Circumstantial evidence for the involvement of strychnine in the Castaing case is buried in the vast memoirs of Alexandre Dumas.

It is not difficult to infer the sources of Dumas's interest in alkaloids. He records a number of friendships with doctors, including the man who treated him after he was wounded in a duel. When it came to the Castaing case, he not only gave a detailed account, but also at the age of 21 he was there in person on the last day of the trial:

> I was present at the final tragedy; I begged a day's holiday from M. Oudard in order to see the end; I was present among the number of those young people whom the condemned man, in a moment of exaltation, of delirium, perhaps, invited to his execution....No, I was not present at the execution; for, I must admit, I could not possibly have borne such a spectacle...at fifteen minutes past two, as the quarter chimed, his head fell....All was ended by half-past two, and those who wished to have comedy after tragedy still had time to go from the Place de Grève to take their stand in the queue outside the Théâtre Français.[11]

He records how Castaing's intense interest in botany and chemistry was evidenced by his heavily annotated notebooks which were produced in evidence.

They attested the determination shown in his researches and the profound study he had made of poisons, of their various kinds, of their effects, of the palpable traces some leave on different bodily organs, while some, quite as deadly and more insidious, kill without leaving any vestiges perceptible to the eyes of the most learned and experienced anatomist.

These poisons are all vegetable poisons: brucine, derived from false angostura; strychnine from Saint Ignatius nut; morphine from pure opium, which is extracted from the Indian poppy.[12]

Dumas's account, although romanticised, is supported by the factual records of the trial itself, in which strychnine gets equal mention alongside morphine. The results of the autopsy, and the symptoms before death, were considered by a further panel of physicians making ten in all, including Vauquelin, Magendie and Orfila. They found signs of "violent cerebral inflammation," or "Arachnitis." The prosecutor said:

> Could all or some of the cadaveric symptoms noticed during the autopsy of Auguste Ballet...have resulted from the employment of a deleterious substance, and notably of emetic, morphine acetate or strychnine?...The ten physicians

unanimously made the following response; "The cerebral congestion [and] other post-mortem phenomena observed in the body of Auguste Ballet, and described in the autopsy report, are very frequently found in the bodies of individuals dead of certain illnesses. *Some poisons, among which we include emetic, morphine acetate and strychnine, can also produce the same changes.*"[13]

Vauquelin and the others were unable to say for certain. There were no available forensic tests for these alkaloids, and some scientists considered that their detection might forever prove an impossibility; furthermore, at this time the symptoms of human poisoning by the alkaloids had hardly been documented. What they were prepared to say was that Ballet's post-mortem signs were not inconsistent with poisoning by emetic and morphine (they do not mention strychnine at this point).

There also is circumstantial evidence for strychnine or brucine in the extreme bitterness of the wine with which Castaing dosed Ballet on the first night. When given a glass of a drink containing too much lemon juice, people ask for more sugar to be added, they do not invite the servant to comment on the taste. Then there were the symptoms during the second night; violent spasms of "nervous shivering," during which Auguste complained that he could not keep himself still. This could have been consistent either with antimony-induced vomiting suppressed with morphine, or with strychnine poisoning. The symptoms are not clearly enough described. Then, more suggestively, there is the stiff neck and periodic spasms not long before he died.

During the endless dissections of the Palmer case decades later, Ballet's symptoms were adduced as evidence that morphine and tartar emetic could produce Cook's convulsions; but, the converse may well be the case. The similarity of their symptoms is consistent with Castaign having got his hands on strychnine. As Alfred Swayne Taylor wrote in 1875:

> It has been stated that morphia in large doses does not operate as a narcotic, but as a stimulant to the nervous centres, causing violent convulsions. In some instances the convulsions are said to have assumed a tetanic character, resembling those caused by strychnia....The statement appears to be based

more on theory than fact. I know of no cases to support the theory, but many averse to it. Such theoretical views become of serious import to medical evidence, when it is pretended that the tetanic symptoms of strychnia are not to be distinguished from those caused by large doses of morphia! They just serve the purpose of unsettling everything and settling nothing. One medical authority has announced that all the symptoms assigned to poisoning by strychnia in Cook's case might be explained by supposing that he had been given three grains of morphia!...If this was poisoning by morphia, then medical experience and observation go for nothing in reference to poisoning by strychnia. Such a theory carries its own refutation.[14]

Leaving aside Taylor's usual polemical tone, the implications are clear. Auguste Ballet's symptoms, so like those of John Parsons Cook three decades later, were not evidence that both had been poisoned with morphine alone, as some claimed. They were evidence that both had been poisoned by strychnine, but the authorities in both countries stifled all mention of the properties of this new and deadly poison after the Castaign case. The sudden onset of convulsions only three or four minutes after the administration of the "medicine" on Auguste's final night is highly indicative of strychnine. On the other hand, the fact that he survived until about noon the next day indicates that strychnine could not have been the only poison used, or else that he eventually succumbed to repeated doses.

Endnotes

1. Thompson, 1925: 148.
2. *Pharmaceutical Journal*, 1856, XV, 289–292.
3. According to his own testimony (Anon, 1823b), his first names were Jules-Samuel, but everywhere else he is called Edmé-Samuel. I cannot find any explanation for this discrepancy.
4. Anon, 1823a; Irving, 1918: 158.
5. She is not named in any of the accounts of the case.
6. There is a great deal of uncertainty about how Ballet described the taste of the wine. The domestic testified that Ballet complained it was "amer," (bitter), but that when she herself tasted it, she had found it "sur" (sour). Castaing denied that Ballet had used the word "amer"

but had instead said that it was "trop acide, trop acerbe" (acid, caustic). Lemon juice is sour or acid, but strychnine (and morphine) are bitter (Anon, 1829a: 116).

7. Irving, 1918.

8. Magendie in 1813 cited a number of cases where people had accidentally or deliberately taken far more than the usual medicinal dose of 2–3 grains. Most of them survived after vomiting but there were occasional fatalities. In a horrible experiment, he gave a large dose to a dog, then tied up its oesophagus to prevent it vomiting. The dog died, although other dogs given larger doses and able to vomit survived (Magendie, 1813:54). There was much discussion of tartar emetic's properties in connection with the famous Bravo poisoning case of 1876. According to one expert witness, "The effects produced by antimony are many and varied, more so than any other poison" (Bridges, 1956:201, 216, 258). Bravo may have accidentally poisoned himself with exactly the same combination, tartar emetic and morphine (laudanum), as discussed here.

9. Chevalier, 1820.

10. Although at first they were administering quinine, which would have had little effect, not morphine (Magendie, 1813).

11. Dumas, 1907–9: II: 463–464.

12. Ibid. 460.

13. Anon, 1823a: 85. Original emphasis.

14. Taylor, 1875: 570.

CHAPTER 9

The Fop, the Scotsman and the Opium-Eater

Though those that are betrayed
Do feel the treason sharply
Yet the traitor
Stands in worse case of woe

Shakespeare, *Cymbeline*, 1609

British authorities had a lax attitude towards the sale and purchase of poisons, in marked contrast to France. It is true that the French authorities found it impracticable to regulate as closely the sale of poisonous substances used in commerce, such as copper arsenite (Scheele's blue). This and other arsenical pigments were used, for example, in wallpaper printing and were responsible for the accidental or deliberate poisoning of many people, including, it is thought by some, Napoleon. But nevertheless, compare this regulatory regime with the report of an 1839 coroner's inquest at Nottingham on the suicide of Ann Burdett, age 16:

> William Yeomans, a grocer, druggist and beer-seller of Charles Street, deposed that he sold poisons as well as other articles, though he was really a pawnbroker by trade. His Wife sold the girl a pennyworth of arsenic, unlabelled, in a screw of paper to dress a bedstead.[1] Mrs. Yeomans testified that she knew nothing of arsenic, she was not aware that the amount she sold Ann Burdett would kill a score of people;

111

"I cannot write at all, neither can I read but very little." She rather thought that the arsenic pot was inscribed in Latin, but when it was produced before the Coroner, it appeared to be labelled "Arsenic" in some sticky substance. This was the second time that the pair had been before the court in similar circumstances, the previous time having led to the death of a baby over a mix-up involving laudanum. The Coroner reprimanded them in the strongest manner. After the reprimand, Yeomans abruptly asked for his arsenic pot to be returned, but the Coroner ordered it destroyed.[2]

In England, the history of strychnine poisoning in the decades before Palmer burst upon the scene is shrouded in mystery. Only one known case, a quarter of a century earlier, illuminates these decades, and that is only a guttering candle in the darkness.

The artist and writer Thomas Griffiths Wainewright (1794–1847), moved in aristocratic circles.[3] To those who had known him, he was "a man of mediocre but varied and serviceable talents, devoured by conceit and a fop in dress and manners." He was a good art critic, championing Turner and Blake when they were unpopular, but in his own work he specialised in "drawings of female beauty, in which the voluptuous trembled on the borders of the indelicate," and his writings, under pseudonyms such as "Janus Weathercock" are highly pretentious, even allowing for the flamboyant style of the day.[4]

Had Wainewright been as agreeable as he seemed on first acquaintance, history would have judged him kindly as a man of second-class talent who did much to bind together the artistic circle of his day. But there were two obstacles in the way of his fulfilling this destiny. Firstly, he did not have the material resources to act the generous host on a grand scale. Secondly, he was much too egotistical to concede the limelight.

Wainewright's maternal grandfather, Ralph Griffiths, had been a wealthy publisher, and the orphan was brought up in his elegant residence, Linden House at Turnham Green just outside London. When the grandfather died, in 1803, he left the house to his son, Wainewright's uncle, and left Thomas himself £5,000 tied up in annuities. This produced a modest income totally inadequate for the lifestyle that Wainewright had marked out for himself. His library contained richly bound volumes on astrology, the occult sciences, and poisons.

In 1823 Wainewright forged the documents relating to the trust fund, allowing him to get his hands on the legacy capital. The forgeries were not detected until much later, when he pointed out, with some justification, that the money was his anyway. A few years afterwards he had spent all of this, his uncle was still alive and he had urgent need of more working capital. On January 16, 1828, George Griffiths became suddenly ill with "convulsive pains in the stomach" and died. Wainewright was disappointed to find that the uncle, who had been only a shadow of the grandfather as a publisher and much preferred gardening, had left only £5,000; sufficient to run Linden House for a while, but nowhere near enough for Wainewright's prodigal scale of entertainment. He began to engage in trivial art frauds, but urgently needed another source of major funding.

The 1820s showed an explosive growth in the life insurance industry. In 1806 there were nine insurance offices in London, but between 1806 and 1826 another thirty opened; twenty of these went bust, for the business was intensely cutthroat. What had started as a means of enabling someone to provide for his family after his death, was now "applied to a variety of purposes."[5] Creditors would insure the lives of their debtors. When there was a marriage settlement, the husband would routinely insure the life of his wife to cover against the capital reverting to someone else if she died. There had to be a legitimate financial interest in insuring someone else's life, however; the courts would rule a policy void if this were not the case.[6]

In 1830, Wainewright's mother-in-law, Mrs. Abercromby, came to live with Wainewright and his wife Eliza, bringing her two other daughters, Helen and Madeleine, Eliza's younger half-sisters. Almost immediately after arriving at Linden House, Helen Abercromby, described as a "healthy, blooming young woman," was taken by Wainewright and Eliza to the Palladium Insurance Company. They insured her life for the sum of £3,000 and paid a single premium of £39. They also visited the offices of the Pelican, the Eagle, the Globe, the Hope, the Imperial and the Alliance. In some of them further cover was granted, but in others they were turned away because the officers had been comparing notes. At one office, Helen said that she intended to travel abroad. At the Alliance, a Mr. Hamilton quizzed her closely, and warned her in the clearest possible terms that she was placing herself in danger by her course of action. Helen seemed blasé, and assured him that she was sure that there was no one who could possibly have the objective of harming her. Before long, Helen's life was insured for a total of £16,000. In the midst of all this,[7] Mrs. Abercromby died suddenly with agonising symptoms resembling those experienced by George Griffiths two and a half years previously. Six days before, she had written a will in favour of her daughter Eliza, Wainewright's wife, but she had very little to leave.

In mid-December of the same year, Helen, sometimes accompanied by Eliza, sometimes alone, embarked on a frantic round of visits to lawyers' offices to transfer the life insurance policies to Wainewright. She also made two wills, one in Madeleine's favour, one in the Wainewrights', dated on consecutive days (Wainewright ill-advisedly showed them to his creditors, saying "If one of them falls, the other will do").

On the night of the fourteenth, a Tuesday, the three of them returned to Conduit Street, where they were temporarily lodging, and had a meal of oysters and bottled beer. Helen, who had got her feet wet during their visit to the theatre, had a restless headache and was violently sick. She remained ill for a week, becoming partially blind, despite Dr. Locock's administration of various remedies including tartar emetic. On the following Tuesday she seemed much better, but shortly after the doctor left at noon, a servant, Harriet Grattan, saw Eliza Wainewright give her some powder in jelly. The doctor later testified that none of the nostrums he had prescribed for her had been in powder form.

Two hours later, while Wainewright and Eliza were out for a walk, Helen began to suffer the symptoms of strychnine poisoning: feelings of dread, then intermittent convulsions separated by intervals in which it appeared that she might recover; she was conscious throughout. During the last interval, Helen told Locock, "Doctor, I thought I was gone to heaven, but you have brought me back to earth." But the elderly nurse, Sarah Handcocks, who had been with the family for decades and had seen both George Griffiths and Mrs. Abercromby die, said aloud how her symptoms were identical with those at the death of her poor mother. The convulsions returned, and Helen died at four o'clock.

Years later, on discussing the matter with a fellow guest at a dinner party hosted by Charles Dickens, Dr. Locock said that his suspicions had been raised very soon by the old nurse's insistence that the circumstances of the three deaths had been identical. But this was with the benefit of hindsight. In the pre-Palmer era, a doctor coming across the first-ever known case of deliberate poisoning by strychnine, only twelve years after its isolation in France, could not be criticised for missing it entirely. The contents of Helen's stomach were given to a Dr. Graham, but he was unable to do anything meaningful with them.

If Wainewright thought that his financial problems were now settled, he was mistaken. The insurance companies refused to pay up; not on the grounds that Helen had been murdered, but on grounds of misrepresentation. Wainewright therefore launched a legal suit against one of them, financed by a loan raised on the security of Linden House and the other policies. He sailed to Boulogne, a common destination of English debtors, to await the successful outcome of his action.

Little is known about his seven-year stay in France. Legends that he poisoned an Englishman and a married Frenchwoman there are probably specious. He may or may not have been imprisoned for three months or so by the Paris police, found to have been in possession of strychnine, but released because he was travelling under an assumed name and was treated as an eccentric Englishman. In 1835, however, while he was in France, the Bank of England discovered the forgery of the trust fund documents up to thirteen years previously, and a warrant for his arrest was issued. Forgery was still a capital offence.

The lawsuit against the insurance company did not reach court until June 1835. The attorney general, Sir John Campbell, later the presiding judge at Palmer's trial, said that Helen's symptoms had been fully consistent with strychnine poisoning, but that only mineral poisons were detectable in the stomach. The judge ruled this inadmissible, and said that if the object was to prove a charge of murder, more cogent evidence would have to be produced; and in any case, a civil court was not the place to bring up such allegations. He said that the legal position was that even if it were proved that the plaintiff had murdered Miss Abercromby, the office would still have to pay up, because Helen Abercromby herself would not have been guilty of any fraud. The case could only be tried firstly, on the question of whether there had been misrepresentations when the policy was taken out, and secondly on whether it had genuinely been for Miss Abercromby's benefit, or really for the benefit of the plaintiff Wainewright. Campbell's comments nevertheless struck home. The jury, "almost petrified" by the mention of strychnine, were unable to agree, and the case was withdrawn after thirteen hours in court.

Undeterred, Wainewright and his associates applied for a retrial, heard six months later.[8] By then, attitudes had hardened. In his opening address, Wainewright's counsel decided to pre-empt the question of possible murder, and unwisely referred to the "foul and unjustifiable aspersions" cast on his client. But Campbell shifted his attack; he did not even refer to the presumed murder. Instead he adduced much more evidence from the insurance offices to strengthen the case for misrepresentation. This time the judge's summing-up was much less favourable to the plaintiff. The judge told the jury that by the actions taken, "Mrs. Wainewright was placed in a situation in which the law would not allow any person to stand—namely that of having an interest in procuring the death of a fellow creature—by unlawful means." The jury decided immediately in favour of the insurance company on both grounds. Linden House was sold.

For some reason which has never been established, Waine-wright returned to England a year and a half afterwards and was arrested. He may have been following a woman. One romantic version has it that one of the actuaries whom Helen Abercromby visited before her death had fallen in love with her, and swore his revenge on her murderer, trailing Wainewright through France and tricking him back to England, perhaps with the aid of a female accomplice.[9] Dickens made this version into a novelette, *Hunted Down*. A slightly more prosaic version of events, but one not without fanciful appeal, has it that Wainewright, while stay-ing incognito at a hotel in Covent Garden, heard a noise outside and pushed the blind aside for an instant. Someone in the street saw him, and cried out, "That's Wainewright, the bank forger!"

In the squalor of Newgate Prison on June 27, 1837, he was unexpectedly seen by a party including Dickens: "My God! There's Wainewright!" looking mean, fierce, and dishevelled. But before long, he began to revel in his notoriety, reconstituting his image as that of an aristocratic chancer who had gambled and lost. "They think I am here for £10,000," he told another visitor, "Even in Newgate I am a gentleman. The prison regulations are that we should each in turn sweep the yard. There are a baker and a sweep here besides myself. They sweep the yard; but, sir, they have never offered me the broom." Dickens based the unpleasant character Rigaud Blandois in *Little Dorrit* on what he saw of Wainewright in Newgate.[10]

Wainewright was tried before Mr. Justice Vaughan and Mr. Baron Alderson, who nearly twenty years later was to sit on the bench at Palmer's trial. At the time of the Wainewright trial, the establishment was under pressure from the media to do away with the capital penalty for forgery, and it served their interests not to push for it. While Wainewright had been in prison, Madeleine Abercromby, now married, had presented two other insurance policies for payment, on the grounds that she had been an innocent party. The insurance companies correctly divined that Wainewright would, out of spite, do anything to ensure that she was not successful. They therefore presented him with a time-worn deal: cooperate with us fully, plead guilty to the lesser charges and we will put a word in for you. (Another version has it that Wainewright wrote to them, offering his services.) But even the lesser charges carried a maximum penalty of transportation for life.

Wainewright was duly convicted and sentenced to transportation to Tasmania. The rest of his time was spent in the uncongenial surroundings of "that metropolis of murderers, university of burglary and all subterhuman abomination, Hobart town." Here he gained a reputation as the most sinister inmate of the penal colony, "dreaded, disliked and shunned by everybody." A fellow prisoner lay dying in the hospital where Wainewright was working as an orderly and post-mortem assistant. Wainewright glided to the man's bedside and hissed into his ear, "You are a dead man. This time tomorrow your soul will be in hell and my arm will be buried up to the elbow in your guts!" During this period of exile he was said to have admitted killing all three of his victims with strychnine and morphia.[11]

It is almost certain that Helen Abercromby was killed with strychnine. But the Wainewrights must have subjected her to a weakening regime, probably involving tartar emetic, to make her eventual death more plausible. It must have caused Wainewright a wry smile that Dr. Locock at one point prescribed the same substance as a curative.

There is no doubt concerning Eliza's complicity. Not only did she apparently administer the final dose to Helen, but she conspired in setting up the various insurance premiums. Since neither she nor Wainewright were ever tried for murder, little is known about her, and the degree of her involvement will never now be known. It has even been speculated that she was the prime mover, and that Wainewright was in her power. But if he was, he soon broke free, for after he left England for France he never saw her again.

What is much more mysterious is Helen's evident connivance in the steps that led to her own death. In view of her comment to one of the insurance offices that she expected to go abroad shortly, it has been speculated that the three of them, Wainewright, Eliza and Helen, and possibly Madeleine too, initially had a conspiracy to defraud the insurers by faking Helen's death abroad. She was then tricked, either because Wainewright and Eliza never intended to do other than poison her, or by their having a change of mind under the pressure of their creditors. Or did Wainewright tell Helen that he was going to elope with her after first disposing of Eliza? This idea has been floated, but it is a gross insult to the memory of an innocent young girl whose brain was taken out and examined after her death to prove Locock's theory that she had died of cerebral congestion as a result of wet feet and eating oysters. A visitor at Newgate asked Wainewright how he could have had the barbarity to kill such a fair and trusting creature as Helen. "Upon my soul, I don't know, unless it was because she had such thick ankles," Wainewright is said to have replied.

There is much less evidence concerning the other two possible murders. They are tied to strychnine only by the evidence of Sarah Handcocks.

However Wainewright learnt about strychnine, it was almost certainly not through books. In the 1820s there was very little literature indeed concerning strychnine or even nux vomica available to the general public. A pleasant little illustrated booklet, written in 1823 to help in cases of accidental poisonings does not even mention it.[12]

Wainewright certainly had some luxuriously bound old books on poisons in the large library at Linden House, but it is not known what they were and there was no trace of them in the auction records when the library was sold.[13] Charles Norman[14] describes him as having had two books on the subject, neither of them old, the titles according to him being "an essay on the action of poisons on the living body" published in 1829, and A Traité de Toxicologie published in 1814. Motion[15] repeats the first title and says it was, "The first such work in English," although there is no such title. The book meant was presumably Christison's A Treatise on Poisons in Relation to Medical Jurisprudence; but by the time this appeared, George Griffiths for one was already dead, and it seems in any case highly incongruous for Wainewright, a literary man and collector of rare books, to rush out and buy a brand new copy of a highly technical scientific treatise the moment it was published. The 1814 title referred to by Norman is presumably Orfila's book, but an 1814 edition could not have referred to strychnine. If Wainewight had used nux vomica powder, he could probably have put the information he needed together from published sources, but the rapid demise of Helen Abercromby after a dose of powder small enough to be given in a jelly argues strongly for pure strychnine, not nux vomica.

In the early 1830s, the poison almost universally used by clandestine killers was the tasteless and undetectable arsenic, and for Wainewright to have used strychnine argues both that he saw poisoning as a kind of sophisticated game, in which he could show himself as out of the common run, and that he had inside-track information on strychnine. If he did indeed use it to kill his relatives, he was in possession of not only strychnine, but also of the facts about strychnine, within a decade of the publication of their results by Pelletier and Caventou. Even if he used it only on Helen, he had discovered the technique of joint poisoning with strychnine and tartar emetic within only a dozen years of strychnine's discovery. This is an impressive performance by someone who is not a doctor. How did he achieve it?

The writer Thomas De Quincey's best-known work is the autobiographical essay, *Confessions of an English Opium-Eater*; but he also wrote another essay, *On Murder Considered as One of the Fine Arts* (1827 and 1839). The subject matter of the *Murder* essay seems to us peculiar and tasteless, and struck many people similarly even when first published. De Quincey described the idea of the essay coming to him while lying awake through lack of opium, and postulates, in the light-hearted callous vein typical of the style of the essay itself, how "some few dozen of useless old women I could frighten out of their wits and this wicked world."

The essay describes a dining club established to appreciate and encourage style in the act of murder. People may question this philosophy, he says, but quotes in aid an incident when his friend Samuel Taylor Coleridge rushed round the corner to see someone's house on fire, and returned disappointed because the fire had not been up to scratch. It was not that Coleridge had a grudge against the householder, or wished to see him suffer; it was merely that he wanted to see a good, satisfying fire. As with fires, so with murders, and thus he develops his theme of connoisseurship.

The *Murder* essay appeared in two parts, twelve years apart. The second part begins with a humorous apologia. All his neighbours had come to hear of his little aesthetic essay, he says, and had assumed that because he was a member of a dining circle, he was responsible for encouraging people to go out and commit murders. The time has come to outface his critics and bring out the rapier to show that he is not discomfited:

> As to murder, I never committed one in my life. It's a well-known thing amongst all my friends. I can get a paper to certify as much, signed by lots of people. Indeed, if you come to that, I doubt whether many people could produce as strong a certificate. Mine would be as big as a breakfast tablecloth.

He was, of course, right. De Quincey was a mild-mannered man, "Singularly gentle in his nature,"[16] who never offered violence to anyone. He did, however, follow every detail of murder cases reported in the newspapers, and when he was editor of the *Westmorland Gazette* in 1818–1819, he filled up every spare column inch with lurid trial reports, together with an account of Dr. Ure's attempt to resurrect the corpse of a murderer using galvanic batteries. His fascination with violence was that of a peaceable dreamer with "a discursive and encyclopedic mind" and a twenty-year opium habit. When the murder essay was republished in 1854, re-edited by De Quincey himself, a lengthy postscript was added giving a fanciful and not particularly entertaining account of the murders (not by poison) committed by John Williams in 1812 and by the Mackean brothers in 1826. But surely there were murders much closer to home about which De Quincey could have written with more authority?

For De Quincey and Wainewright were at the very least acquainted.[17] Richard Woodhouse records that on December 6, 1821, "I dined at Taylor and Hessey's this day, in company with the Opium-Eater, Reynolds, Lamb, Cunningham, Rice, Wood, Wainwright [*sic*] and Talfourd." Everyone was in good spirits but De Quincey, who was quiet. This is their only recorded meeting; Wainewright issued a return invitation a few days later, but De Quincey turned it down. De Quincey disliked Wainewright, not because he thought him a possible murderer, but because he thought him a shallow fop, trying to steal the limelight from writers of real genius like Hazlitt.

Also missing from the 1854 version of the *Murder* essay is a lengthy passage in which De Quincey refers to a fictional nephew, who wants to become a member of the club but who was:

> Certainly born to be hanged, and would have been so long ago but for my restraining voice. He is horribly ambitious, and thinks himself a man of cultivated taste in most branches of murder, whereas, in fact, he has not one idea on the subject but such as he has stolen from me....If we were to elect him, why the next thing we should hear of would be some vile butcherly murder... "John," said I ... "believe me, it is not necessary to a man's respectability that he should commit a murder...." John made no answer; he looked very sulky at the moment, and I am in high hopes that I have saved a near relative from making a fool of himself by attempting what is as much beyond his capacity as an epic poem. Others, however, tell me that he is meditating a revenge upon me and the whole club.

The most likely scenario is that on first publishing Part 2 in 1839, with Wainewright's trial and deportation a recent event,

De Quincey faced off possible accusations of bad taste with this counterattack of deliberate and considered bad taste. Then in 1854 with Wainewright a distant and embarrassing memory, he quietly dropped it.

During the 1820s, De Quincey was dividing his time between London, Grasmere in the Lake District, where he intermittently rented Wordsworth's cottage, and Edinburgh. While staying in the Scottish capital in December 1820, he wrote to his wife, "Invitations crowd upon me so fast that I hardly know how I shall get through all of my writing &tc."[14] At this time, Robert Christison, almost a contemporary of Wainewright's, was in Paris learning about strychnine a few months before his appointment to the chair of medical jurisprudence in Edinburgh. A biography of De Quincey[18] specifically mentions Christison as one of the Edinburgh figures whom he knew, but much firmer links than this can be traced.

In 1830, Christison wrote an account of his involvement in the Earl of Mar insurance case.[19] This concerned the Earl's death in September 1828 at the age of 57, two years after he had taken out life insurance and without disclosing to the insurance company a long-standing opium habit. The company contested the claim and Christison was consulted as an expert witness. Christison's account of his evidence includes descriptions of several cases of opium use "which had been communicated by a friend." Case 3 is "a well-known literary gentleman who has taken laudanum with some intermissions for twenty years, and...has now attained his forty-fifth year. He is spare in form, looks older than he is, but is capable of undergoing a good deal of bodily fatigue, and enjoys tolerably good health so long as he takes sufficient exercise." This is clearly De Quincey, who was born in 1785, and for whom the description is precise. He and Christison were intimates, for "in Case 3, I can state from personal observation that even after being told of the habit existing, no-one could discover it from the gentleman's appearance, conversation or acts." They had even discussed bowel movements; "the subject of Case 3 merely requires a little rhubarb on average once a week, as he finds that exercise counteracts any constipating effect the opium may have." Christison's Case 8, an "eminent literary gentleman" about sixty years of age and on a quart of laudanum a day, is evidently Coleridge (born

1772). "Enormous as this dose may appear, I am assured the fact is well known to his acquaintances," Christison says. Coleridge was a close friend of De Quincey, and it is clear that the "Friend" referred to by Christison, who had put the professor in touch with a number of confirmed opium addicts, and who is described himself as "Case 3," is none other than De Quincey himself.

It seems highly probable that De Quincey learned of the astounding properties of strychnine from Christison. It is not difficult to visualise this ever-curious and discursive man discussing with his London friends what he has heard in Edinburgh about the newly discovered and remarkable poison, and the sly Wainewright vowing to show them a thing or two, poisoning George Griffiths within a few months of the appearance of the first part of *Murder*. De Quincey remarked much later that he was disappointed with himself that he had not realised Wainewright's nature, and it seems overwhelmingly likely that their acquaintance was not based on just the single recorded meeting. The Castaing poisoning, which took place the year after the recorded dinner party of 1821, may well have featured in their conversation.

The Wainewright case was suppressed after his deportation and even before his death. The prospect that detailed knowledge about the techniques of strychnine poisoning might spread abroad was clearly an unsettling one. The bulk of Wainewright's papers found their way into the hands of the novelist Edward Bulwer Lytton who used them as source material for one of his tedious novels, after which they were destroyed.[20] Virtually all material relating to the first half of the career of Sir Robert Christison also seems to have disappeared. The fact that De Quincey wrote at length in the 1850s about other killings about which he had no personal knowledge, but that neither he nor Christison ever referred to the Wainewright case, is circumstantial evidence for their horrified realisation that they had been indirectly involved in three murders.

Endnotes

1. i.e., To kill bugs and fleas.
2. *Lancet*, 1839: 254.
3. Curling, 1938; Motion, 2000.

4. Hazlitt, 1880. The fact that Wainewright was only a third-rate writer is supported by the fact that some other pieces under the pseudonym "Egomet Bonmot," previously attributed to him, have recently been reattributed to another, completely unknown, person (see Marc Vaulbert de Chantilly in Myers, Harris and Mandelbrote, 2001:111–142).

5. Francis, 1853.

6. In 1809 Sir Mark Sykes, while drunk in the course of a dinner party, offered to pay anyone a guinea for every day that Napoleon Bonaparte survived, in exchange for a premium of 100 guineas. Three years later, having shelled out over 1000 guineas, he refused to pay any more and the policy-holder, a clergyman, sued him. The court ruled in Sir Mark's favour, since the clergyman had no legitimate financial interest in Napoleon's life (Francis, 1853).

7. Actually in August 1829 not August 1830 as previously reported (Motion, 2000: 156).

8. The reports of the trials are reproduced at Hazlitt, 1880: 337–355. These reports do not give Counsel's opening address, in which strychnine was mentioned to the horror of the jury.

9. Hazlitt, 1880, and especially Thornbury, 1879, were the source of most of the dubious anecdotes about Wainewright; for example, that he went about with a signet ring full of strychnine. Thornbury is highly unreliable; to take just one example, he states that strychnine is tasteless.

10. For further suggestions about the influence of the Wainewright case (and also Palmer) on Dickens, and on Bulwer Lytton, see Altick, 1970: 72, 128–129.

11. Andrew Motion (Motion, 2000:171 especially) is an apologist for Wainewright much as Robert Graves was for Palmer. The case does not hang together, though.

12. Anon, 1823.

13. Marc Vaulbert de Chantilly, *Thomas Griffiths Wainewright and the Griffiths Family Library*, in Myers, Harris and Mandelbrote, 2001:111–142 (p. 123 for the fact cited).

14. Norman, 1956.

15. Motion 2000:171.

16. Hogg,1895: Eaton, 1936; Lindop, 1981.

17. Hazlitt (Hazlitt, 1880:lxxiv (footnote)) speculates whether Wainwright might have give De Quincey the idea for the Murder essay, and/or suggestions about what to put in it. See also Motion 2000:119.

18. Masson, 1914:87.

19. *Edinburgh Medical and Surgical J.*, 1832:37:123.

20. Motion, 2000:284–285.

CHAPTER **10**

It Will Be the Test-Tube and the Retort That Will Hang Him

There is a wonderful sympathy and freemasonry among horsy men. Be one of them, and you will know all that there is to know.

Sir Arthur Conan Doyle, *A Scandal in Bohemia*
(The Adventures of Sherlock Holmes)

One day in 1857, the prime minister received an unusual delegation at no. 10 Downing Street. The mayor and other citizens of Rugeley in Staffordshire, about 150 miles northwest of London, together with their member of parliament, Mr. Alderman Sidney, asked Lord Palmerston to arrange for their town to be renamed. The name of Rugeley in their opinion had become so sullied by a recent notorious poisoning case that it was no longer usable.[1]

After listening to their representations, Palmerston realised that he was being presented with an opportunity to let loose a shaft of wit at the expense of the unfortunate provincials. He told them that they could have their name change, provided that the replacement name should be "Palmerston." This rather brilliant sally allowed him to indulge in a little piece of mock egoism, knowing full well that it was an offer that they could never take up.

For the man who had brought such disgrace to this hitherto obscure little market town was called William Palmer.[2]

The short career of William Palmer is a prime example of what a really wicked fellow had become able to accomplish by the mid-nineteenth century, given the support of an efficient railway system, a network of lesser scoundrels, and ready supplies of strychnine. He could bet all over the Midlands. He could get to London and back in a day, so as to fall into the clutches of moneylenders such as Mr. Thomas Pratt of Queen Street, Mayfair, a solicitor and front man for various people investing their money in Palmer and his friends at an annual rate of 60%. Now, thanks to the industrial revolution and the growth in middle-class prosperity, families of obscure origin like the Palmers were able to dispose of amounts of money that no one short of a duke would have been acquainted with a few decades before. They also took out life insurance policies.

The Palmers of Rugeley were wealthy. Palmer's father Joseph had been a "coarse, unscrupulous, insolent, pushing" timber merchant, who in partnership with a man called Hodson (who was sleeping with Palmer's wife, Sarah) defrauded the Marquis of Anglesey, an old soldier who as Lord Uxbridge had famously lost a leg at Waterloo. The two men sold off the trees ("those excrescences of nature, given by Providence to pay the debts of gentlemen") on Uxbridge's estate and pocketed most of the money. Local rumour had it that Hodson was actually William Palmer's father.

On Joseph Senior's sudden death at the dinner table in 1837 he left no less than £70,000. Even "Superior Dosset" Forsyte, the fictional founder of the Forsyte dynasty in Galsworthy's *Forsyte Saga* left only £30,000 in 1850; in 1830 the field labourers of southern England had rioted to try to obtain a wage of half a crown (25¢) per day. Of this vast sum, each of Joseph's seven children, of whom William was the fourth, received £7,000, the remainder going to Sarah Palmer. A condition of her inheritance was that she should not marry again.

William Palmer was cunning rather than clever. As a child, he was notable for his apparent kindness masking a sly manner, and for his interest in vivisection experiments. This evidently influenced his choice of career, for he first became apprenticed to a firm of Liverpool druggists where he escaped punishment for stealing cheques when his mother bought the firm off. He then studied medicine at Apothecaries Hall, and promised John

Steggall fifty guineas to get him through the exams, but never paid up[3] and set up in Rugeley as a surgeon. Soon, however, he lapsed the practice and devoted himself to racehorses; becoming a "turfite" in the terminology of the Attorney General at his trial. (The bankruptcy writ against him described him insultingly as "surgeon, apothecary, dealer and chapman" [itinerant peddler].) He was shortish and plump, with a face that belied no trace of wickedness. He was said to have been a compulsive nail-biter and hand-washer. While a medical student, he was rumoured to have fathered no fewer than 14 illegitimate children.

In 1847 he married Annie Brookes, a ward in chancery and the illegitimate daughter of a Colonel Brookes late of the Indian army, by his ill-tempered and promiscuous housekeeper Mary Thornton, whom Palmer subsequently probably poisoned. Brookes was the youngest of five brothers, all of whom committed suicide. When Colonel Brookes shot himself, he left considerable property to his daughter, but as his estate went into chancery she required the consent of the trustees before marrying. These trustees, two local men, knew quite a lot about Palmer and opposed the match; considerable perspicacity was necessary on his part to overcome their objections and Annie's initial reluctance.

The Palmers lived in good style, as well they might quite legitimately, but it was the betting that tightened the screw on William's fortunes. Within only a few years he was in debt to the tune of five figures. When Annie died in September 1854 heavily insured, the insurance offices paid up, despite mumblings about an enquiry. Palmer next insured the life of his brother Walter, a notorious drunk; he asked for cover for a colossal £82,000, but had to settle for £13,000 because all of the offices except one had been tipped off about Annie. This policy became lodged with Palmer's moneylender Pratt as security for large amounts of credit. Palmer spent much of the money on two racehorses of his own, Nettle and The Chicken,[4] which failed to deliver the goods. Nettle fell while lying second in the Oaks, breaking the jockey's leg. By this time, rumours were circulating that Palmer had poisoned his wife, and Nettle may have been nobbled by the anti-Palmer faction.[5] In the meantime, Palmer was making sure that Walter got all the drink he wanted; innkeeper John Burgess recalled having supplied the house with nineteen gallons of gin between February and July 1855.

When Walter died in August 1855, the insurance office refused to pay up and negotiations about the policy dragged on, bringing Palmer's financial crisis to a head; on November 6, 1855, Pratt issued a writ for £2,000 but agreed to hold it over while Palmer desperately tried to raise the money. Palmer also owed large sums to a certain Padwick, and his house and chattels were mortgaged to a Birmingham solicitor, Edwin Wright, for £10,400.[6] Another writ was taken out by Pratt against Palmer's mother, for other debts had been guaranteed by her. Or rather, she had not guaranteed them, despite what Mr. Pratt thought, because Palmer had forged his mother's signature on the bills.

One George Bate, a failed farmer employed by Palmer as a kind of estate manager, living in rented accommodation at a shilling a week, had a lucky escape. Palmer tried to insure him for £10,000, describing him to the insurance office as a gentleman farmer with the finest wine cellar in Staffordshire. The office sent an investigator and Bate was discovered digging turnips. Another person who no doubt read the accounts of Palmer's subsequent trial with horrified fascination was Jane Bergen. At the time of Walter's death she was sitting on a cache of 34 letters from Palmer, in one of which he mentions procuring an abortion for her. They were negotiating terms for their return; she was blackmailing Palmer while he was in full spate as a murderer. Jane was the daughter of the chief superintendent of the Stafford Rural Constabulary. Palmer and his mistress Eliza Tharm (his late wife's former maid) in turn used the fact that Jane Bergen had demanded payment for an abortion to blackmail the police.[7]

Next off the starting gate, and far less fortunate, was John Parsons Cook, for whose murder Palmer was tried and executed.

Cook's good fortune was his misfortune. His mare Polestar was entered for the Shrewsbury Handicap on November 13, 1855. He and Palmer were in the habit of going to race meetings together. Like Palmer, he was a small-town professional man who had abandoned his vocation, in Cook's case the law, better to concentrate on horses. He was not robust, but was in reasonable health for a mid-Victorian, in an era when virtually everybody had rotten teeth and a variety of other chronic infections. Cook was a mild hypochondriac and was worried that he had syphilis; he had been taking mercury for it. However, the post-mortem showed only a very small old healed syphilitic ulcer or chancre on his penis, and no other signs of the disease. He also thought that he had a serious throat illness, and told a friend that when he had eaten a joke nut containing cayenne pepper which someone had given him on the platform at Liverpool station, it had nearly killed him. But a doctor who examined him in London on several occasions testified that he had seen only mouth ulcers and inflamed tonsils, and had prescribed harmless medicines in place of the mercurial pills. Cook was 28 years old, played cricket and hunted three times a week.

November the thirteenth at Shrewsbury was cold and wet underfoot, but Cook was in good spirits and became even more so when Polestar won at seven to one, raising him enough in prize money and successful bets, about £2,000 in all, to clear his own relatively modest debts. These, it later transpired, included a charge against Cook's two nags Polestar and Sirius, which Cook (actually Palmer) had mortgaged with Pratt for £440 in the form of a cheque for £375 and a warrant for £65 made out "to order" and sent to Doncaster post office on Palmer's instructions. Pratt made the advance against Cook's two horses without ever having met him or having seen his handwriting, receiving through the post an acceptance bearing a forged signature, which, of course, he was unable to recognise as such. Palmer intercepted the cheque at Doncaster, wrote Cook's signature and used the money to repay one of his own bills to Pratt. The defence argued that these were joint debts between Palmer and Cook, and that forging the signature was just an irregularity, not uncommon in such louche circles.

Cook's illness began the following evening at the Raven Hotel in Shrewsbury where he and Palmer were staying. The two of them, together with two other layabouts, George Myatt, saddler from Rugeley, and Ishmael Fisher, described as a "sporting wine merchant," of Shoe Lane, London, were drinking brandy and water. Cook offered Palmer some more grog, and Palmer said he would not have any more until Cook had drunk his.

Cook drained most of his brandy and said, "There is something in it; it burns my throat dreadfully." Palmer picked up the glass, tasted the small amount remaining, and replied, "There is nothing in it." A man called Reid came into the room and Palmer handed him the glass and asked him if he thought there was anything in it. Reid and the others said it was impossible to say as the glass was by now empty. Cook left the room and was ill. He gave Fisher £700 to look after and told him that he thought Palmer had been "dosing" him, then vomited. He was very ill all that night and a doctor was sent for, but was somewhat better in the morning.

The following evening, a Mrs. Anne Brooks—who described herself as being "in the habit of attending race-meetings," although "my Husband does not sanction my going when he knows about it"—went to see Cook in his room. In the passageway outside, she encountered Palmer standing at a small table, holding a glass tumbler containing some liquid up to the light. He said, "I will be with you presently," and carried the glass into his sitting room. He then came out and gave her a glass of brandy and water, which did her no harm. Cross-examined, she revealed that a great number of people had been unwell with sickness and purging at the Shrewsbury meet, and there had been talk in the town of the water supply being poisoned. (It is not inconceivable that Palmer could have poisoned a number of people nonfatally to provide him with an alibi for Cook's deterioration.) Mrs. Brooks seems to have been a doughty witness.

Counsel—Are you intimate with Palmer?

Mrs. Brooks—What do you mean by intimate?

Myatt's evidence at the trial contradicted the others' statement of events at the Raven Hotel, and supported Mrs. Brooks' statement about the outbreak of illness. Myatt said that Palmer himself had vomited out of the window of the fly taking them back from Stafford station to Rugeley. When cross-examined at the trial he was subjected to character assassination and made to admit that he was a lifelong friend of Palmer and had visited him for two hours at Stafford gaol.

Cook and Palmer returned to Rugeley on the evening of November 15th, Chicken having lost in its race earlier in the day. Cook continued his residence at the Talbot Arms, almost opposite Palmer's house. Nothing of note occurred on the 16th, but on the evening of Saturday the 17th, the pace of events began to quicken.

The key witness for the prosecution was Elizabeth Mills, chambermaid at the Talbot Arms. In the morning Palmer ordered a cup of coffee for Cook, which Mills gave to Cook in his bedroom. Palmer was in the room at the time. She did not see Cook drink it, but about half an hour afterwards found that the coffee had been vomited. During the day she saw Palmer perhaps four or five times in Cook's room and later heard him say that he would send over some broth. She did not see it brought over, but saw it in the kitchen. Palmer's charwoman testified that she had fetched it from the Albion, another inn a short distance away. Palmer had poured it into a cup in his own kitchen and told Mrs. Rowley to take it to Cook saying that it had been sent up by Jeremiah Smith, a solicitor whom we will meet later.

Elizabeth Mills met Palmer going up the stairs to Cook's room. He was anxious that Cook should take the broth. About an hour and a half afterwards she found that it had been vomited. In the morning Cook said he felt more comfortable. On the Sunday a large breakfast cup of broth was brought to the Talbot Arms and Mills took some of it up to Cook's room. She tasted about two tablespoons. About half an hour afterwards it made her very sick, and she vomited violently all the afternoon, about twenty times in all.

Over this weekend Cook was attended by Dr. Bamford, an 80-year-old doctor who had already obliged with death certificates for Anne and Walter Palmer and several of Palmer's children. Bamford, who had been in practice since 1803 or before, was unqualified. The Medical Registration Bill did not become law until later, and even then, unqualified doctors then practising were allowed to continue. At first, Palmer told Bamford that Cook was ill owing to excessive drink on the Friday night, but Bamford did not agree because Cook told him that he had only drunk two glasses of wine. Bamford prescribed pills containing calomel (mercurous chloride), morphia, and rhubarb.

On Sunday evening Cook was better again and in good spirits. On the Monday morning Mills took him up a cup of coffee. He did not vomit that, got up about lunchtime and dressed and shaved himself. About eight o'clock that night the housekeeper gave Elizabeth Mills a pillbox to take upstairs to Mr. Cook's room, which she placed on the dressing table. It was wrapped up in white paper and she did not determine whether it contained pills or not. These were probably more pills sent in by Bamford, but the prosecution case was that Palmer substituted them with some of his own. After she had placed the pillbox on Cook's dressing table Palmer went in, and she saw him sitting down by the fire between nine and ten o'clock. Soon afterwards she went to bed.

At about a quarter to twelve another servant, Lavinia Barnes, called her up. She heard a noise of violent screaming from Cook's room while she was dressing. She found him sitting up in bed, asking for Palmer and beating the bedclothes, with both his arms and hands stretched out. He said, "I cannot lie down. I shall suffocate if I do. Oh, fetch Mr. Palmer." His body, hand and neck were moving—a sort of jumping or jerking. "Sometimes he would throw back his head upon the pillow, and then he would raise himself up again." He appeared to have great difficulty breathing. His eyeballs were much projected. He called aloud, "Murder," twice. He asked her to rub his hand; she found it stiff and paralysed. He recognised Palmer when he came in, and said, "Oh, Doctor, I shall die." Palmer replied, "Oh, my lad, you won't." Palmer then left and returned in a few minutes with some pills. Cook said he could not swallow them. At Palmer's request she gave Cook a teaspoonful of toast and water. He snapped at the spoon, so that it was fast between his teeth. While this was going on, the water went down his throat and washed the pills down. Palmer then gave him a draught from a wineglass of dark, heavy-looking liquid which he vomited up immediately. The stuff he vomited smelt, she thought, like opium. Palmer said that he hoped the pills were not returned, and he searched for the pills with a quill. He said, "I cannot find the pills." After this, Cook appeared to be more at ease.

Lavinia Barnes also saw Cook in great agony at about midnight. According to her recollection, Palmer told him, "Don't be alarmed, my lad." Another housekeeper said that when she went up to Cook's room between eleven and twelve o'clock that night she found him on his own, and he seemed disappointed that she was not Palmer. He told her that it was Palmer he wanted. The next morning Cook, who was again better, asked Elizabeth Mills if she had ever seen anyone suffer such agony as he had been in during the night, and she said no, and asked him what he thought the cause was; he told her it was the pills that Palmer had given him.

On the Tuesday William Jones, Cook's family friend, and another doctor, arrived. The three doctors, Jones, Bamford and Palmer, had a consultation in Cook's presence, and he told them that he did not want any more pills because he objected to morphia. The three then met without him and agreed that he should take some more of Dr. Bamford's pills. Bamford went back to his surgery to fetch some and was surprised when Palmer accompanied him and insisted that Bamford write the directions on the box. Palmer took them back to the hotel, but with a delay. He was careful to draw Jones's attention to the writing on the box, saying how remarkable it was for an 80-year-old man. Cook initially refused them, but gave in to pressure and took them at about half-past ten. He vomited at first, but retained the pills, and lay down to sleep. Ten minutes afterwards he started up, said that his neck was stiffening.

Elizabeth Mills had stayed in the kitchen through concern for Cook and was alerted by the bell from his room ringing just before midnight. She went up and found him sitting up in bed supported by Jones. He asked for Palmer, and she went over the road to fetch him. She rang the bell at the surgery door and Palmer came to the window; she could not see if he was dressed or not. He arrived at the hotel only two or three minutes later, and on coming into Cook's room his first remark was that he had "never got dressed so quickly in his life." Mills waited outside the room with Lavinia Barnes, and shortly afterwards Palmer came out. She said that Cook was much the same as he had been the previous night, and Palmer told them, no, he was not so ill by the fiftieth part. They waited again outside the door and heard Cook ask to be turned onto his side. Then he died, quietly.

Jones said that the body was so arched up that had he placed it on a level surface, it would have rested on the head and the heels. He thought, and continued to think, that Cook had died of tetanus. At the trial there was extensive cross-examination of a Rugeley widow who had a sideline laying out bodies with the help of her sister-in-law; she had laid out many corpses; she could not exactly say how many, but many children; but she was unable to give clear answers as to how much stiffer Cook's body had been than she would normally have expected. One of the hotel staff, though, told a reporter that he had "lay quite curved; all of a ruck, like," and they had had to use cords to tie him down.

Few people have ever seriously contended that Palmer did not poison Cook at all, but the exact details of when precisely, and with what precisely, are very uncertain. There are many similarities with the Castaing case more than thirty years before. Apart from all the doings involving hotel rooms, intolerably bitter drinks, doctors getting up in the middle of the night and so on, the same three poisons, tartar emetic, morphine and strychnine come into play, at least potentially. Robert Graves[8] argued that Palmer gave Cook tartar emetic as a kind of practical joke, not believing that it could kill.

The evidence concerning Palmer's possession of strychnine is tainted. Charles Newton, another scoundrel, was assistant at Rugeley to a surgeon, Mr. Salt. Palmer called at the surgery on Monday November 19th at nine p.m. and got three grains of strychnine, for which Newton did not charge him anything. This was after the initial "vomiting" poisoning of Cook, and just before the onset of his "tetanic" phase. However, Newton did not mention this gift at the inquest on Cook; he said that this was because Salt and Palmer were enemies, and Salt would have been angry with him for selling or giving anything to Palmer. Palmer would normally have got all his medicines from Mr. Thirlby, who had taken over Palmer's practice. Incidentally, Newton was Thirlby's illegitimate son.

The next day, Newton said he went into the shop of Hawkins, a druggist, and saw Palmer there. As soon as Newton entered the shop, Palmer steered him out into the street, saying that he had something to discuss with him, but the matter that Palmer raised was of no significance. While they were talking, someone came up to talk to Newton, and Palmer took the opportunity to slip back into the shop. Newton's evidence on this occasion was corroborated by Charles Roberts, the apprentice to Mr. Hawkins, who said that Palmer had initially bought two drachms of prussic acid, and then when he sidled back into the shop he asked for six grains of strychnine and two drachms of Batley's solution of opium. While Roberts was making up the order, Palmer had stood with his back to him, looking into the street. It had been two years since Palmer had bought anything from them. No written record was made of the purchase.

At the trial, the attorney general put forward two possible explanations for the fact that Cook suffered severe tetanic poisoning on two successive nights. One was that Palmer had deliberately given Cook a sublethal dose the first night in order to bolster the explanation that he was suffering from recurrent paralysis. The alternative was that the first sample of strychnine had been substandard and consisted mainly of brucine.[9]

Palmer had reckoned without the energetic William Stevens, Cook's stepfather, who arrived in Rugeley on November 23rd and immediately became suspicious. As a result and at his own expense he engaged the forensic scientist Dr. Alfred Swayne Taylor. Stevens, a retired merchant living in London, had last seen Cook there on November 5th, when he had remarked on Cook's healthy appearance. This was the last he saw of him before he learnt of his death. When he arrived at Rugeley, Palmer, whom he had met only once before, took him to view the body at the Talbot Arms. Stevens was greatly struck by the appearance of the countenance; the tightness of the face muscles fixing it in the *Risus sardonicus* so typical of strychnine poisoning.

He quizzed Palmer about Cook's finances and was told that everything was owed to him, Palmer, including the racehorses which were mortgaged; he had the papers to prove it. Stevens brought up the question of burial and Palmer indicated that if that was all that Stevens was worried about, he would see to it himself. Stevens replied that as executor, he had the responsibility and that he intended to bury Cook in London in his mother's grave. Palmer said that would be of no consequence as long as the body was fastened up at once. Stevens asked him for the name of a respectable undertaker in Rugeley and Palmer astonished him by revealing that he had already ordered a strong oak coffin and liner, or shell. Stevens said he had no authority to do so. Before Stevens left for London, he asked Jones to go and look for Cook's betting book, but it was nowhere to be found. Palmer told him that it was of no manner of use if he found it, and Stevens replied that he was the best judge of that, and that the book must be found. Palmer replied in a much quieter voice, "Oh, it will be found, no doubt."

The very next day, by the two o'clock train from London, Mr. Stevens was back bearing a letter from his solicitor to Mr. Gardiner, the solicitor at Rugeley. Another passenger on the train was none other than Palmer. An extraordinary series of conversations took place in a linear trajectory between London and Rugeley, for in the 1850s the trains had neither restaurant cars nor lavatories, and stopped every fifty miles or so for the passengers to eat, drink and relieve themselves. At Euston, Palmer explained his presence on the train by explaining that he had been summoned to London the night before by telegraph. In the refreshment room at Wolverton, Stevens said that there ought to be a post-mortem, an opinion that Palmer appeared to accept with equanimity. At Rugby, Stevens told him that he intended to consult a solicitor about his stepson's affairs. On arrival at Rugeley, Palmer offered to introduce Stevens to a suitable local solicitor, since he knew them all intimately. Stevens deliberately changed his demeanour and said sternly, "Mr. Palmer, if I should call in a solicitor to give me advice, I suppose you will have no objection to answer him any questions he might choose to put to you?" According to Stevens, "He replied, with a spasmodic affection of the throat, which was perfectly evident, 'Oh, no, certainly not'." The moon was shining, but Stevens could not swear to the fact that he saw Palmer's face distinctly. Later that evening they met again in the coffee room of the Talbot Arms. Palmer told him that it was a very unpleasant affair to him about the bills by which he was owed money. Stevens replied that he had heard a different account of Cook's affairs, according to which they could only be settled in Chancery, to which Palmer said, "Oh, indeed," in a low tone. Stevens announced his intention of taking a solicitor along to Hednesford, where Cook's horses were kept. Palmer advised him against it, and Stevens told him that he would be the best judge of that. When asked if he had attended Cook medically, Palmer said no. He asked Stevens if he knew who would perform the post-mortem. Stevens said he did not.

The first post-mortem on Cook was carried out in a room at the Talbot Arms on November 26th. The previous evening, Palmer sent for Newton, and according to Newton, asked him how much strychnia would kill a dog, and whether it would be found in the stomach. Newton says he told him that about a grain would suffice, would not cause inflammation, and would not be detected. Newton said that he thought Palmer then said, "It is all right," as if speaking to himself, and snapped his fingers.

The following morning, Newton met up with Palmer on the way to the autopsy, and as they walked to the Talbot Arms. Palmer primed him about Cook's preexisting syphilis. He offered Newton some brandy. "It will be a stiff job," he told him, without apparent irony.

The farcical events in Rugeley on November 26th, in this case where there was strong suspicion of foul play, almost defy belief. The dissection was carried out by a Mr. Devonshire, an undergraduate of London University with no practical experience, assisted by Newton and supervised by Dr. John Harland of Stafford, who turned up without any instruments. (He testified that he later examined some white spots on the surface of Cook's stomach by scraping them with his finger; "I should have examined them with a lens if I had had one.") Palmer, who joined him on the way to the Talbot Arms, offered to lend him some instruments; they were old acquaintances. Palmer said he was glad that Harland was doing it, as they might have sent someone Palmer didn't know. Harland said that he had heard there was a suspicion of poisoning, and Palmer replied, "Oh, no, I think not; he had an epileptic fit on Monday and Tuesday night, and you will find an old disease in the heart and in the head." He also told Harland that a "queer old man" was accusing him of poisoning Cook and stealing his betting book, which was of no use to anyone.

The dissection took place in the presence of "several people"; no one seems to have counted them all at the time. The body was exceptionally stiff for someone who had been dead for six days. Harland described what happened next:

When the intestines and stomach were being placed in the jar, and while Mr. Devonshire was opening the stomach, I noticed Palmer pushed Mr. Newton onto Mr. Devonshire, and he shook a portion of the contents into the body. I thought a joke was passing among them, and I said, "Do not do that" to the whole. Palmer was smiling at the time. After this interruption, the opening of the stomach proceeded. It contained about, I should think, 2 or 3 ounces of brownish liquid. It was stated that there was nothing particular found in the stomach, and Palmer remarked to Mr. Bamford, "They will not hang us yet." The stomach was then emptied into the jar along with the stomach itself. [*Devonshire testified that at this point "I punctured the anterior surface of the stomach, and a spoonful of the contents fell out on the chair."*] I then tied the jar over with two bladders and sealed it, and placed it on the table beside the body. At that time Palmer was moving about the room. My attention had been called away by the examination, and I missed the jar for a few minutes. I called out, "Where is the jar?", and Palmer, from the other end of the room, said, "It is here; I thought it more convenient for you to take it away." Palmer was standing a yard or two from a door at that end of the room. [*A plan of the room was produced, and it was shown that the door beside which Palmer was standing was not the one by which they had entered, and did not lead anywhere useful.*] I got the jar from him. I found there was a cut, hardly an inch long, through both bladders. The cut was quite clean, as if nothing had passed through. I asked who had done this, and Palmer, Mr. Devonshire, and Mr. Newton all seemed to say they had not done it. I told Palmer I should take the jar to Mr. Frere. He said, "I would rather you take it with you to Stafford, if you would take it there," but I took it to Mr. Frere's house, tied and sealed in the way I have told...when I returned to the Talbot Arms, Palmer asked me what I had done with the jar. I said I had left it with Mr. Frere, and that it would go to either London or Birmingham that night for examination.

A charitable explanation of these farcical goings-on might have been that Harland had been well briefed about the suspicions surrounding Palmer, and gave him enough rope literally to hang himself by letting him be present. But it appears much more likely that he was just exceedingly incompetent. No proper report about the observations or findings, signed by those present, was ever written.

There were two other post-mortems: one a couple of days later to remove further organs, and again later, when the potential significance of the state of the spinal cord was realised, a further exhumation to examine it. It had now started to disintegrate, so not a lot was learned, but there were no obvious disease symptoms.

Palmer's interventions in his own case had not reached their apogee yet, however. John Myatt, the postboy at the Talbot Arms, testified that Palmer had enquired whom he was engaged to drive to Stafford station in the fly on the morning of November 28th. When Myatt told him that it was Mr. Stevens along with the jars, Palmer said that there was a £10 note for him if he would upset them. Counsel asked if it were not true that Palmer had actually said, "I should not mind giving £10 to break Mr. Stevens' neck,"[10] but the postboy did not agree. Palmer had told him that it was a humbugging concern and that Stevens was a suspicious, troublesome fellow. Myatt declined the offer.

Not stopping there, Palmer called upon the services of his former schoolfellow Samuel Cheshire, who was now Rugeley postmaster. The wretched Cheshire was persuaded to open a letter addressed to Mr. Gardiner, the solicitor, of Landor, Gardiner and Landor in Rugeley, an act for which he received a prison sentence of one year for interfering with the mails. This letter contained the results of the analysis of Cook's stomach contents. Palmer wrote to the coroner, Mr. Ward, sending him a letter by hand of Bate, who was instructed to deliver it so as not to let anyone see what he was doing. The letter enclosed £10 and gave Ward, who was later brought to book for venality and incompetence, various helpful hints for his conduct of the enquiry. He also sent Ward two hampers of game.

The irony of all this hyperactivity of Palmer's is hard to escape. His transparently obvious ruses only served to incriminate him and to fill the courtroom with a succession of witnesses providing more and more circumstantial evidence as to his guilt. He would have done himself far more good by just lying low.

For the analysis of Cook's stomach contents, communicated in the letter from Professor Alfred Swaine Taylor of Guy's hospital to Mr. Gardiner, had failed to reveal any strychnine.

Endnotes

1. The status of this anecdote is uncertain. According to some, it is a myth. It is too good to leave out, however.
2. Bennett, 1856; Anon, 1856; Knott, 1912; Fletcher, 1925; Lewis, 2003; *Lancet*, 1856, 563–566: *British Medical Journal*, 1856, 429, 461; *Illustrated Times* special issues 2nd February and 27th May 1856. The latter source gives the most entertaining details, while the *Lancet* account is the most accurate report of the medicolegal issues.
3. *Pharmaceutical Journal*, 1856:XVI:5–11. Palmer obviously reckoned that if Steggall had bent the rules to allow him to pass, he would be in no position to recover the debt.
4. Sold after Palmer's execution for 800 guineas and renamed by its new owner "Vengeance" (Altick, 1970: 160). Another of the stable was bought at the auction by Prince Albert.
5. An old Yorkshire trainer was heard to observe at Epsom, "Hoi's noa going to win Oaks as hoi's poison'd woife" (*Illustrated Times*).
6. The account in the Notable British Trials series, edited by a barrister (Knott, 1912), plays up Padwick, describing him as a "notorious racing man and moneylender," makes light of Pratt's involvement, and does not even mention the money owing to Wright. The legal profession looking after its own. The *Times* at the time of the trial tore into Pratt, however (22 May 1856:8). At the inquest on Walter Palmer, Pratt broke down, screaming excitedly, "How can you ask such questions of a man with three young children and a wife who will probably be ruined by this affair?" At the Old Bailey he was described as "a tall and sinister figure with enormous brown whiskers almost meeting under his chin, but the face of a small boy and the voice of a retiring female." He impressed with the cold, merciless manner in which he gave evidence but shortly afterwards he became raving mad and died in an asylum. (Fletcher, 1925).
7. Palmer's biographer, Fletcher, thought the letters highly disgusting, but he was a Victorian. To our taste they begin mostly harmless enough, with some coded references to sexual encounters (such as where Palmer says he will meet up and perform "an operation" on her). But when the subject of abortion comes up, they turn sinister. A proposed abortionist will be "as silent as death." It is highly probable that Palmer was trying to insure her life and would have killed her. The full text of the letters has now been published (Lewis, 2003: 167–178).
8. Graves, 1957. Robert Graves's book is a romantic, factional account of Palmer's life that contains much background material. But Graves's central thesis that Palmer just happened to be surrounded by people dropping dead, fails to convince.
9. The *Pharmaceutical Journal* (1856:XVI:8) revealingly opined that "In a small shop in a country village it is more than probable that it would be otherwise [than pure]."
10. In other words, just expressing general disgust with Stevens and not bribing Myatt to upset the fly.

Figure 1. Traditional medicine in India. Hakeem (physician) with servant carrying medicines, 1856.

Figure 2. *Strychnos nux-vomica*, after M.A. Burnett, 1847.

Figure 3. Strychnine; (left) Nux vomica seeds bought over the Internet in 2002 from a supplier in India. A kilogram, containing enough strychnine to poison perhaps 200 people, cost $30 plus postage. (right) A sample of Parke Davis & Co strychnine sulphate for injection (circa 1900) bought over the Internet by the author in 2002 as a collector's item. The total content of the pills is approximately half a grain, possibly a fatal dose. Neither package was queried by UK customs.

Figure 4. Medicine in the early nineteenth century. *Taking an Emetic*, by Cruickshank, 1800. Emetics were a mainstay of medicine in this era. Incorrect ideas about them probably led to the development of Dr. Castaing's method of disposing of the Ballet brothers.

Figure 5. Eight leading French doctors of the early 1800s. Pierre-Eloi Fouquier, who was the first to introduce nux vomica seriously into therapy, is at top right. The front row is (l. to r.) Mathieu Orfila, the founder of forensic science, Guillaume Dupuytren, who found tetanus mysterious, and François Magendie, who showed how strychnine could kill animals.

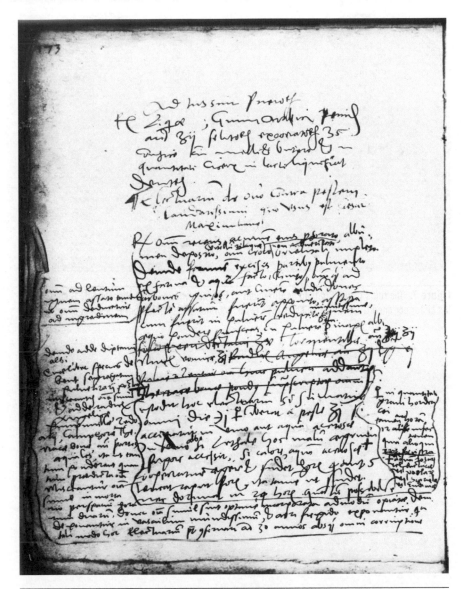

Figure 6. Knowledge of nux vomica reaches the West. Part of a manuscript in the Vienna state archives, ascribed to the physician Georg Handschius (Prague, 1550). The following can be made out: "Electuarium de ovo laudatisssimum contra pestem quo usus est Caesar Maximilianus" and the words "Nux Vomica."

Figure 7. The ruins of the Chateau de Bitremont in Belgium, where Count Hippolyte de Bocarmé and his sister poisoned Gustave Fougnies in 1851 (with nicotine). The chateau burnt down in about 1988.

Auguste Ballet
MORT a St Cloud le 1er Juin 1823

Hyppolite Ballet
MORT a Paris le 5 Octobre. 1822

Castaing
Docteur en Medecine
Condamne a mort
Dessiné d'apres Nature

Figure 8. Edmé-Samuel Castaing and his two victims, Auguste and Hippolyte Ballet.

Head of a Convict, very characteristic of low cunning & X revenge !

Figure 9. A circumstantial chain. Sir Robert Christison; Thomas De Quincey; Thomas Wainewright. Who told what to whom?

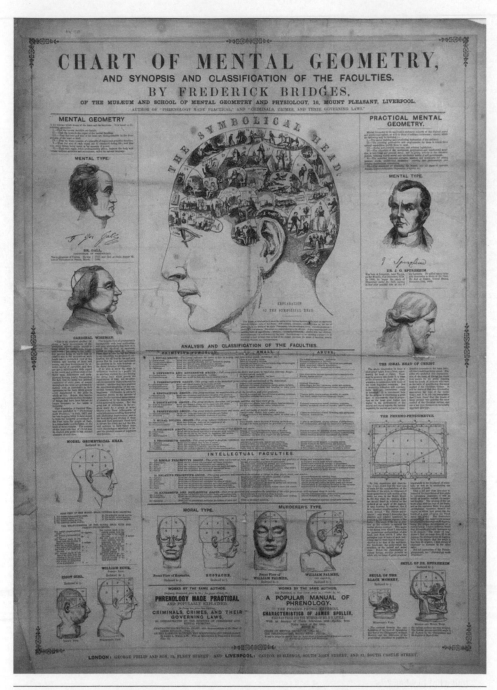

Figure 10. A Victorian phrenology chart incorporating William Palmer as the criminal archetype.

Figure 11. A victim of tetanus, showing the characteristic arching of the back (opisthotonos) and other muscle contraction.

Figure 12. The Old Town Hall, Rugeley, during the inquest on Walter Palmer, 1856. Taylor is giving evidence, the corrupt coroner Ward is in the chair, and the ancient Dr. Bamford stands towards the right. Palmer refused to attend. (From the *Illustrated Times,* February 1856.)

Figure 13. "The strychnine must be in here somewhere." Alfred Swaine Taylor and his colleague Dr. Rees.

Figure 14. Christiana Edmunds.

Figure 15. Thomas Cream.

Figure 16. 1904 Olympic marathon winner Thomas Hicks under the influence of strychnine and alcohol.

TOO LITERAL BY HALF.

Scene.—*A "cheap" chop-house not a hundred miles from L—nd—n.*

Waiter. "Paysir? Yessir—Whataveyeradsir?"
Matter-of-fact old gentleman (who has been reading the "Quarterly" on *"Food and its adulterations.")* "Had? why, let me see: I've had some horsetail soup, spiced with red-lead and shop-sweepings: a plate of roast cow, and cabbage boiled with verdigris: a crust of plaster of Paris, baked with alum and bone-dust: half-a-pint of porter brewed from quassia and strychnine: and a cup of charred liver, annatto, and other unknown ingredients."
[*Exit Waiter for a Straight-Waistcoat, and a Stomach-Pump*

Figure 17. Who says there was no strychnine in beer? Cartoon from *Punch*, 1855.

Figure 18. Jean Pierre Vaquier and Mabel Jones.

Figure 19. Robert Robinson and R.B. Woodward.

Figure 20. The 1917 assassination plot. The three female conspirators in prison. Left to right; a wardress, Hettie Wheeldon, Winnie Mason, Alice Wheeldon. (From the *Illustrated London News,* Feb. 10, 1917.)

Figure 21. Transvestite cabaret artiste Molly Strychnine with her accompanist Fuckoffsky.

CHAPTER **11**

Shaken in Every Possible Way

In antimony, great though his faith,
The quantity found being small
Taylor's faith in strychnine was greater
For of that he found nothing at all

Anon. Quoted by Robert Graves,
***They Hanged My Saintly Billy,* 1957**

Palmer's trial was notable for the number of medical and forensic witnesses called. The staggering total of 103 medical men were subpoenaed, and a couple of dozen actually gave evidence. Amongst the questions that had to be grappled with were the following: were Cook's symptoms consistent only with strychnine poisoning, or was there some other possible medical explanation? Notably, could his symptoms have been consistent with the disease tetanus? Had the three post-mortems revealed any signs attributable to some fatal disease? Why had strychnine not been detected in the stomach contents; how quickly did it disappear from the body before and after death? The trial served only to point up the current lack of scientific knowledge about the alkaloid. Palmer did not give evidence. The accused was not entitled to do so until the Criminal Evidence Act of 1898.

The very concept of an expert witness was essentially an alien one to the society of the 1800s, and took a long time to become fully recognised. Emergence of science as an important human activity was a challenge to the balance of power between the existing professional groups. In 1821, it was ruled that a man of science, "without recognised learning" and basing his arguments on the novel process of experimentation, had no more status than a mechanic. This decision settled for many years the status of chemists, despite the fact that they frequently made large sums of money as expert witnesses in litigation.[1] The same experts tended to be called again and again; a sort of cartel system operated, by which virtually all "accepted" experts appeared for the prosecution, and there was a gentleman's agreement that they would not appear for the defence in another case. Those scientists who went along with this status quo acquired vast prestige and a sometimes undeserved reputation for omniscience. They were consulted again and again and had to spread themselves pretty thinly. Thus, Henry Letheby was medical officer for health, physician, analytical chemist, professor of chemistry, food analyst, and chief inspector of illuminating gas for the city of London.

In general, the relationship between science and the law throughout the Victorian era was at best an uneasy one, as we shall see during the trial. An indication of scientific ignorance among the senior judiciary comes from this exchange. Serjeant Shee for the defence, cross-examining one of the medics (Morley), tried to ask him, "What are the component parts of strychnia?" and was interrupted by the bench.[2]

Mr. Baron Alderson: You will find that in any Cyclopaedia, brother Shee.

Before his trial, Palmer's debts had been foreclosed and his possessions put up for sale. By the merest chance, Inspector Crisp of the Rugeley police was able to intercept Palmer's medical textbooks and found among them his copy of Steggall's *Manual for Students Preparing for Examination at Apothecaries Hall* with marginal annotations on the effects of strychnine. In the meantime, Palmer was taken to London from Stafford jail on January 20, 1856 to appear before Mr. Justice Erle in an action brought by Padwick to recover £2,000 from one of the forged bills. In a bravado display of unprincipled cynicism, Palmer swore on oath that his mother's signature on the bill had been forged by Anne Palmer, now deceased, the wife whom he had murdered, and the action failed because the perpetrator of the forgery was dead.

Over the Easter weekend, 1856, as the nation breathlessly awaited the opening of Palmer's trial, Rugeley was thronged with visitors from all parts of the country, come to inspect the scenes of the undoubted crimes. A facetious bystander was reported as claiming that Palmer's house was to be reopened as an inn, to be named "The Strychnine Arms."[3] The Old Bailey eventually opened its doors in May 1856. The evidence was given before Lord Chief Justice Campbell and two other judges and lasted twelve days, including Saturday sittings. The courtroom had to be enlarged and there were doubts whether it would be ready in time. The case was followed across the Continent, copies of the London papers carrying a full transcript causing daily astonishment in Paris and Berlin. It also made legal history; it was the first ever to be moved away from the local assizes because of the intensity of the feeling generated in Staffordshire, and to allow this a special act of parliament (the Palmer Act) had to be passed.

When the trial eventually opened, the attorney general set the scene in his opening address:

> You have heard, I dare say, of the vegetable product known by the term of nux vomica. In that nut or bean there resides a fatal poison capable of being extracted from it by the skill and operations of man; of this the most minute quantity is fatal to animal life...the poison to which I am referring affects the voluntary action of the muscles of the body, leaving altogether unimpaired the consciousness ...it produces the most intense excitement of all these muscles, violent convulsions take place...by this means respiration is prevented and death necessarily ensues.

These spasms, he said, are known by the medical term "tetanus"; the defence will try to confound the symptoms of strychnine poisoning with the natural disease of tetanus, but he will prove otherwise, and called his witnesses.

Mr. Thomas Blizzard Curling, surgeon at the London Hospital, said that there were two kinds of tetanus, traumatic tetanus resulting from a wound, and idiopathic tetanus which arose "as it were, as a primary disease." He had never seen a case of the latter. The course of tetanus was normally from one to four days with a gradual onset, and he had never heard of a case where the patient would be suddenly stricken with severe symptoms like Cook's followed by an equally sudden respite. He thought that Cook's symptoms were wholly consistent with nux vomica poisoning. Tetanus might arise from something as trivial as a tooth extraction or a fishbone in the throat.

It is here that Victorian ignorance of the causes of disease first begins to show itself. For although Dr. Curling had written a book about tetanus, he thought that the symptoms were essentially due to irritation of the nervous system. This is true in a sense, but he knew nothing of *Clostridium tetani*. The doctors in this pre-Pasteur era could only explain the onset of disease from such trivial causes in terms of a hypersensitive predisposition which some patients might show; a kind of allergy. This ignorance was of great help to the defence, who played on the probable state of Cook's mind as it reeled from despair to elation when his horse won, and the consequent irritation to his nervous system.[4] The evidence about "idiopathic tetanus" was essentially accurate, from an observational point of view, although the attempted explanations were wrong. Tetanus may arise without apparent injury; in fact, the more insignificant the initial wound by which bacteria are introduced into the body, the more likely it is to be ignored and forgotten.[5]

The next witness, Dr. Robert Todd, had seen two cases of idiopathic tetanus. "The proximate effect of tetanus, whether caused by idiopathic or traumatic tetanus, or strychnia, is probably the same on the nerves leading from the spine. The particular affection of the nerves is unknown. ...I have no doubt the peculiar irritation of the nerves in tetanus is identical with the peculiar irritation of the nerves in strychnine poisoning." But like Curling, he opined that the course of Cook's affliction was consistent with strychnine, not tetanus. Sir Benjamin Brodie, Queen Victoria's surgeon, was called to agree with all of this. He handled his cross-examination in a patrician manner:

Mr. Sergeant Shee: Considering how rare tetanus is, would you think that the description of a chambermaid and of a provincial medical man, who had only seen one case of tetanus, could be relied upon by you as to what the disease observed was?

Sir Benjamin Brodie: I must say I thought the description very clearly given.

Later prosecution witnesses dealt with the effects of strychnine in deliberate and accidental poisoning. Chief among these were several people who had been present at the death of the unfortunate Mrs. Sergison Smith, but three other cases were gone into. Witnesses travelled from Glasgow to describe the sufferings of Agnes Sennett, who took an hour to die after taking some strychnine pills intended for another patient at the Glasgow Royal Infirmary. "We were obliged to cut her clothes off because she never moved. She was as stiff as a poker," one of them said. Mr. Edward Moore described how some fifteen years previously, he had nearly killed a Mr. Clutterbuck whom he had been treating for paralysis with strychnia. After Moore increased the dose to a quarter of a grain, he found the patient "stiffened in every limb...screaming, frequently requesting that we should turn him, move him and rub him." He was suffering for about three hours altogether, but the treatment was evidently successful on the whole, because "he was completely recovered the next day after the attack, and the patient himself said he thought his paralysis was better."

The final case was more sinister. This concerned the poisoning of a woman, Mrs. Dove, who could not be named in court. Her maid, Jane Witham, described how she first complained of her back hurting. Her ankles were twisted, her eyes drawn aside and staring. She complained of a pricking in her legs and twitching of the muscles in her hands, which she compared to galvanic shocks. She initially asked for her arms and legs to be rubbed, but later on she could not bear it; if touched between the spasms, another attack would be brought on. She could swallow all the time except on the last day, when her mouth became tightly shut. She took five days to die.

Nearly the whole of the fifth day was taken up with the evidence of Dr. Taylor and of Professor Robert Christison.

Alfred Swaine Taylor was born at Northfleet, Kent, in 1806.[6] After medical studies in London, he attended lectures by Orfila in Paris. He developed an interest in forensic science as a result of treating gunshot wounds during the Paris insurrection of 1830, but after his return to Guy's hospital in London, poisons came to dominate his interests. He was a good teacher and an asset to the profession, but was a poor analyst. His career as a forensic witness was therefore a chequered one.[7]

Taylor had been the chief expert witness at the inquest on Cook. This was under the direction of the Col. Ward, whom Palmer had bribed, and had been as big a bear garden as the post-mortem. Taylor sat next to the coroner and asked more questions than he did. During the proceedings, Taylor said that he had initially thought that Cook had died from antimony poisoning, but having heard the evidence of the witnesses, he was now certain that it had been strychnine. This immediately sparked a highly prejudicial public debate in the press.

Taylor had to admit to never having witnessed the effects of strychnia on a human subject, but had carried out some animal experiments: "I have found half a grain sufficient to destroy the life of a rabbit. Sometimes the rabbits died in the midst of spasms, often with a shriek; in one or two cases they died quietly between spasms. Often when one died in a spasm it was possible to hold it out horizontally by the hind legs for a whole week afterwards without it folding up."

He then went on to describe the colour tests that were used for the detection of strychnine, which he considered very fallacious. Vegetable poisons were much more difficult to detect than mineral ones. Significantly, experiments with his colleague Dr. Rees on the bodies of the rabbits had given mixed results, and in two cases they had been unable to detect any strychnine. His theory for this was that after being absorbed into the bloodstream and travelling to the sites of action, strychnine was changed by the very act of poisoning into undetectable products. If only the minimum dose capable of causing death was administered, there would be none left. This novel theory (which is essentially wrong), was based on experiments carried out after Taylor had failed to find any strychnine in Cook, and was embroiled in controversy through the columns of *The Lancet* and the *Illustrated Times*. It could therefore be inferred, and was indeed implied by the trial judge, that Taylor had something to prove. Cook's viscera, sent to Taylor from Rugeley in the notorious jar, he had analysed for various poisons but had failed to find anything except a trace of antimony. In his opinion, Cook's vomiting at Shrewsbury and Rugeley was fully consistent with administration of tartar emetic (antimony potassium tartrate). This was colourless and tasteless and could easily have been administered in the broth.[8] He complained petulantly about the condition of the sample:

> The part which we had to operate upon was in the most unfavourable condition for finding strychnia if it had been there. The stomach had been completely cut from end to end; all the contents were gone…in journeying up to London it must have been shaken in every possible way.

It was impossible to say with any certainty how long before death antimony had been administered, but the longest period that had elapsed in a poisoning case for antimony to remain detectable in the blood was eight days. He knew of no disease that could mimic the symptoms of strychnine poisoning.

Taylor was cross-examined on what he meant by "traces of antimony." In statements that ring painfully on the ears of modern-day scientists, he manfully attempted to be more quantitative. A trace was a "very small quantity." Did it mean an imponderable quantity? "I do not apply it in that shape. Some chemists mean that. I mean we obtained some quantity in that sense from many parts, and that the quantity thus calculated would make a ponderable quantity in the whole. We have about half a grain." "You did not actually ascertain it to amount to half a grain?" "No. I do not think a quarter of a grain would have explained the quantity that we obtained. I will undertake to say there was half a grain to the best of my judgement." Half a grain of antimony was not enough to kill, and he had changed his mind about the cause of death since he had written to Mr. Gardiner some time ago attributing Cook's death to antimony poisoning.

Professor Robert Christison[9] was effectively the founder of forensic science in Britain. He had written the book A Treatise on Poisons, which was the first English-language textbook on the subject. His experience of strychnine was wide; he had seen experiments done not only on rabbits, but also on dogs, cats, frogs and a wild boar.[10]

The first symptom that he observed in strychnine poisoning was invariably tremor and an unwillingness to move, then the full tetanic symptoms would come on. There was sometimes a quiet interval during their appearance. Making strychnine up into pills containing insoluble resin would delay its absorption, and this technique would be well within the grasp of an ordinary medical practitioner.

"Is there, in your opinion any marked difference between what I may call natural tetanus and the tetanus of strychnia?" he was asked. "I would not rest much upon the little difference of particular symptoms, but rather upon the course and the general circumstances attending them. First, that in all the natural forms of tetanus the symptoms begin and advance much more slowly, and, secondly, they prove fatal much more slowly."

"Now, of the two classes of tetanus, to which should you refer the spasm and other symptoms spoken to by the witnesses?" "To strychnia, or one of the natural poisons containing it: nux vomica, St. Ignatius's bean, snakewood, and a poison called exhetwick.[11] They belong to different plants of the same genus, from all of which strychnia may be obtained. There is no natural disease that I have ever seen or that I otherwise know to which I can refer these symptoms. Where death has taken place from strychnia I should not expect to find it where the quantity taken is small, but where there is considerable excess over the quantity necessary to destroy life by absorption I should expect to find it. Colouring tests are, I think uncertain in some respects. Vegetable poisons are generally more difficult to detect. There is one I know for which there is no test I know of." Christison went on to concur with Taylor in his opinions about the unreliability of the forensic colour tests, and to agree that the contents of the jar were so unsatisfactory that Taylor could not be criticised for not detecting any strychnine.

A final prosecution witness was a physician who had spent many years in India where tetanus was common. He claimed to have seen over forty cases of idiopathic tetanus; he had never seen a case in which the disease ran its course in twenty minutes or half an hour. It did not occur to the defence to point out that nux vomica grows wild in India, and to ask him how he could be certain that all forty of his cases were due to idiopathic tetanus and none to strychnine poisoning?[12]

Lord Chief Justice Campbell was dusting off his black cap. He had a reputation for playing to the gallery. The court usher, who knew him of old, later said that he was certain that Campbell wanted to hang Palmer from the moment that he invited him to be seated with exaggerated politeness.[13] Halfway through the trial, and largely because of all the background material that the jury knew about Palmer's other murders, his goose appeared to be largely cooked. Only a virtuoso performance by the defending counsel, Serjeant Shee, could pull it out of the fire.

At this point, Palmer passed a note to Shee that said, "I wish there was 2½ grains of Strychnine in old Campbell's acidulated draught."

The whole of the seventh day of Palmer's trial was taken up by the eight-hour speech of the chief counsel for the defence. Even by Victorian standards, this was an extremely florid presentation. His opening remarks give the tone of the dissertation:

> May it please your lordships, gentlemen of the jury—I should pity the man who could rise to perform the task which it is now my duty to attempt unoppressed by an overwhelming sense of diffidence and of apprehension. Once only before has it fallen to my lot to defend a fellow-creature upon trial for his life; it is a position, even if the effort should last but for a day, of a nature to disturb the coolest temperament and try the strongest nerves; how much more so when, during six long days, in the eye of my unhappy client, I have been standing between him and the scaffold; conscious that the least error of judgement on my part might consign him to a murderer's doom, and that through the whole time I have had to breast a storm of public prejudice such as has never before imperilled the calm administration of justice!

After further referring to the fact that the trial had been moved from Stafford because of the unlikelihood of his obtaining a fair trial there, he made certain excuses in advance by pointing out that he was the second choice for defence attorney because of the illness of Mr. Serjeant Wilkins. He was actually the fifth choice at least for this tricky assignment.[14]

Hardly an auspicious start to the defence. Shee goes on to portray himself as a man with a burning sense of injustice battling on behalf of his client against almost insuperable odds; a theatrical device that he may have overdone. Later on he was rebuked by the judge for giving the jury the assurance of his personal conviction of the innocence of his client, an assertion which took him beyond the accepted rules of evidence.

Shee first spent some time trying to show that Palmer had nothing to gain from Cook's death. As he put it, would it have been in Palmer's interest that in November 1855 Mr. Cook should be killed in a railway accident? If it was not, then there was no reason for Palmer to have poisoned him. In order to prove this, Shee had to show that Palmer's best hope for staving off his creditors was to rely on further advances from his friend. He played down the seriousness of Palmer's financial position, pointing out that as long as they were reasonably sure of the security of their capital, there was no better place for Pratt and his associates to invest their money than with Palmer and Cook at an annual rate of 60%. But all this was circumstantial, and when Shee tried to flesh out these claims by closely considering the mass of financial documentation, it probably served only to confuse the jury and to convince them even more of the complexity of the mare's nest that Palmer had got himself into.

Shee then turned to the forensic evidence. This put him on the horns of a considerable dilemma. To summarise what Taylor had said: he had been unable to find any strychnine, but the reason was that the dose administered had found its way into the spine, done its deadly work and been used up by the time he had got hold of the unsatisfactory stomach. The problem for Serjeant Shee was that if he praised Taylor, then his client was guilty because the jury would accept Taylor's opinion about why he hadn't been able to find any strychnine. If, on the other hand, he attacked Taylor, Palmer would be guilty because Taylor's inability to find any strychnine was clearly the work of an incompetent.

Shee did his best with this unpromising material, first taking it as read that there had been no strychnine administered because Taylor had not found any, then attacking Taylor for his wild surmises about the fate of strychnine in the body. He emphasised Taylor's lack of experience with human subjects and pointed out that his knowledge of strychnine was based—good, humane man!—on having poisoned just five rabbits some twenty-three years ago, and a few more since the case had begun. A comparably impressive roll of medicolegal experts would give evidence for the defence.

But everywhere he turned, Shee was forced into contradicting himself. He said that Palmer, as a medical man, would not have seriously considered poisoning Cook with strychnine because he would have known that the horrible symptoms could not be confused with anything else. Yet a main plank of his case was that Cook's symptoms had been perfectly compatible with idiopathic tetanus, or if not that with apoplexy, or epilepsy, or at any rate something or other. He makes the most of Cook's real and imagined illnesses at the time of his death, in a passage that is a masterpiece of ambiguous hypocrisy:

> The appearances that were presented at the death of Cook were such as might have been expected by those who had been acquainted with his course of life and his general health, his pursuits—it is a pity to say anything hard of him—his vices—I will not say any more than this—his vices, and the company, the drinking, idle, racing company which he kept...the tonsils of his throat—one of them was very nearly gone, the other was very much reduced in size...he had traces about his person that have been so often referred to, the result of disease, that they need not be more particularly mentioned than they have been already, as to the extent of which and the character of which some little doubt exists; but they did not come by an ordinary and chaste mode of life, you may depend upon it; and altogether, as far as it went, he seems to have been about as loose a young man as one is in the habit of meeting, without being utterly lost to all sense of honour and propriety, which I do not mean to suggest that he was.

Shee then goes on, with much justification, to criticise Taylor, who had first asserted the cause of death to have been antimony poisoning, but had publicly revised his opinion once he heard the evidence of Elizabeth Mills and others concerning the presumed administration of strychnine. He points out that part of the evidence that had changed Taylor's opinion had been the fact that Palmer had bought strychnine on Tuesday, November 19th, but this could not have accounted for Cook's symptoms on Monday, November 18th. He then expands upon the mysteries of sudden death as seen from the theistic viewpoint of the mid-nineteenth century:

> Of all the works of God, the one best calculated to fill us with wonder and admiration, and convince us of our dependence on our Maker, and the utter nothingness of ourselves, is the mortal coil in which we live and breathe....We know in a sense—we suppose—that the soft medullary substance which is within the cavity of the head is the seat of thought, of sensation, and of will...from this medullary substance proceed an infinite variety of nerves, the conduits of sensation from all parts of the body to the soul, and of muscles connected and dependent on them, the instruments of voluntary motion...

Apparently, though, God could slip up from time to time:

> Sometimes, however, these nerves and muscles depart from their normal character, and instead of being the mere instruments of the will of the soul, become irregular, convulsive, tumultuary, vindicating to themselves a sort of independent vitality, totally regardless of the authority to which they are ordinarily subject. When thrown into this state of irritation and excitement their effects are known by the general name of convulsions...

At this point he does not say in detail what his experts' alternative medical explanations for Cook's death will be; thus keeping the prosecution in the dark until the last possible moment.

> My learned friends may tell me, if you venture to impeach the authority of a man like Dr. Taylor...it is incumbent upon you to suggest some other theory of the cause of Cook's death.... I say I am not called on to do any such thing. ...I am not bound to suggest any theory upon the subject.

In other words he will throw out all manner of possible explanations, and if the jury are satisfied that any one of them (and as far as he is concerned, it need not be the same explanation for each member of the jury) might have killed Cook, then Palmer is innocent.

He expands upon the likely ecstatic state of Cook's mind after having won such an amount at the racecourse, and having said that he has said all he intends to about the state of Cook's health, immediately plays one or two angle shots to bring back into play Cook's supposed moral and physical degeneration:

> Conceive him to be a man with right feelings—and it is not because a man falls into the ways of promiscuous licentiousness that he is devoid of all honourable feeling—conceive him to be an honourable man...instead of being known as a levanter and a blackleg, driven from all honourable society. The effect of this is that for three minutes he cannot speak...

He gets what mileage he can out of the drinking that night:

> What in ordinary parlance is called a champagne dinner is a good luxurious entertainment, in which there is no stint and not much self-restraint. I do not mean to say he was drunk. The evidence is he rose from the table not drunk, and therefore it is not for me to say, and the evidence will not justify me in saying, he was.

And goes on to paint a picture, ludicrous to our ears and very possibly to those of the jury too, of a man disintegrating from guilt, drink, dissipation, horseracing and the effects of wet feet virtually exploding in a paroxysm of nervous apoplexy. He says that although he would be incapable of speaking disrespectfully of such as Dr. Todd and Sir Benjamin Brodie, they were hospital surgeons who did not have much experience of witnessing

> ...the class of convulsions which constantly attack people in their own residences in the middle of the night—those convulsions which heads of families and brothers and sisters are most anxious to conceal from anybody but the medical man—those convulsions the known existence of which deprives a young woman of the hope, or a young man of the hope, of marriage.

...raising the spectre of venereal disease, and preying on the medical ignorance and fears of the jury.

He accuses Stevens of conducting a vendetta against Palmer by saying, "You will find the conduct and deportment of the latter were such as would make some men almost kick him," and says that when Palmer asked the postboy to "upset them" while driving the fly to Stafford, he could not have been referring to the jar, as the Crown proposed, because there was only one jar. Palmer was referring to Stevens and his companion because he was so justifiably outraged about the slurs on his character that he wanted to teach Stevens a lesson for accusing him of stealing a worthless betting book, when he had been Cook's best friend and had looked after him in his final illness.

He is more convincing on the subject of Newton. The full statement of what Newton was going to say was not revealed to the Crown barristers, and thence passed on to the defence, until the first day of the trial. Why had he not given his full story in good time for the preparation of the trial papers? Why had he said even less at the inquest?[15]

If Newton was to be believed, Shee says, Palmer obtained strychnine from two different suppliers in this tiny town, including one he was on bad terms with. Why did he need to do this when his first purchase would have been more than sufficient to kill anyone? If on the other hand he had a legitimate reason for buying strychnine, which he did (to poison dogs), there was nothing sinister about him making two purchases. (It was also rumoured that Palmer used strychnine to nobble horses for betting purposes, but this was considered unspeakably base in Victorian times and would have damaged the case.) If Palmer's motives were clandestine, why buy strychnine in Rugeley anyway, when he had just come back from London? Palmer had most certainly been in London at three o'clock that day, and could not have got back to Rugeley until a quarter past ten, whereas Newton said that he had given him the strychnine at nine o'clock.

Why should Palmer, a doctor whose possessions included a copy of John Steggall's book annotated with pencil notes on the effects of strychnine, engage in a conversation with this "stupid fellow" Newton in which he asked him what its effects were? However, we have seen the quality of Steggall's courses of instruction.

Shee is less convincing about Palmer himself. He describes his actions during the days at Shrewsbury and Rugeley, and tells the jury that these cannot be the actions of a guilty man; for example in sending for Cook's friend Dr. Jones:

> He brings a medical man into the room and makes him lie within a few inches of the sick man's bed, that he may be startled by his terrific shrieks, and gaze upon those agonising

convulsions which indicate the fatal potency of poison! Can you believe it?

Yes, perhaps they can, because the prosecution have said that all this was clearly done to cover appearances. This hypothesis has to be nailed with vigour:

Done to cover appearances! No, no, no! You cannot believe it—it is not in human nature—it cannot be true—you cannot find him guilty—you dare not find him guilty on the supposition of its truth—the country will not stand by you if you believe it to be true—you will be impeached before the whole world if you say that it is true—I believe in my conscience that it is false, because, consistently with the laws that govern human nature, it cannot possibly be true.

He finishes on a pathetic note with a character reference for his client taken from a letter to his dead wife, written several years previously:

My dearest Annie: I snatch a moment to write to your dear, dear little self. I need scarcely say the principal inducement I have to work is the desire of getting my studies finished, so as to be able to press your dear little form in my arms. With best, best love, believe me, Dearest Annie, your own William.

Even contemporary commentators found this ludicrous and lacking in taste, not least because nearly everybody in the court-room knew that Palmer had poisoned Annie and slept with the servant, Eliza Tharm, on the night she died.

Shee now called his medical witnesses, who were to provide his alternative explanations for Cook's death. Almost immediately they were given a hard time by the bench.

Thomas Nunneley said that in his opinion, Cook had syphilis, lung disease, throat disease, granules on the spine[16] and led an irregular life. There was a loss of substance from the penis. He had a delicate constitution and both his parents had died young. Excitement and exposure to wet and cold might have brought on convulsive disease. He understands that there are forms of epileptic convulsions in which the patient retains his consciousness. Had he ever met with any? No, he had not. But he had read that they could be brought about by such as indigestion, worms in children.... There was a great deal of depression at Rugeley, he says, as well as over-excitement.

In his opinion, Cook's symptoms were not consistent with strychnine poisoning. He had never had any difficulty in detecting strychnine in the tissues of dead animals, even when dug up 43 days after death in a state of perfect putridity. He is dismissive of Taylor's hypothesis that strychnine is destroyed in the act of poisoning. Cook's full consciousness and his request to be turned over just before he died showed that he could not have been poisoned. Counsel points out that Mrs. Sergison Smith talked throughout and begged to have water thrown over her, then asked to have her legs stretched just before she died. "Does that shake your faith?" he was asked. "Yes," he replied.

The next witness was William Herepath. His appearance at the trial was the result of an extraordinary series of events that took place in Bristol. Simmons, the Bristol magistrate's clerk, wrote anonymously to the treasury reporting that at a recent meeting of the Improvement Committee, Herepath had said that Taylor did not have the talent to find strychnine and that if he, Herepath, had been in charge of the forensic investigations, it would certainly have been found. Simmons was traced through the postmark on the letter and interviewed, and as a result Herepath was subpoenaed.[17]

Herepath had poisoned numerous animals and agreed that it was always possible to detect strychnine in the remains, even when the body was reduced to nothing more than a dry powder. Unmixed with organic matter, he was able to dissolve a tenth of a grain in a gallon of water and detect strychnine in each drop; when admixed with animal tissue, he could detect it in the thirty-second part of the liver of a dog. "Judging from reports in newspapers, I have said in conversation that strychnia had been given, and that if it was there, Professor Taylor ought to have found it."

Dr. Henry Letheby also attacked Taylor's analytical skills: "I do not agree with Dr. Taylor that the colouring tests for the discovery of strychnine are fallacious. They always succeeded with me." He said that Cook's symptoms, including his ability to sit up and ring the bell, were not consistent with strychnine poisoning, and said that some peculiarity of the spinal cord not visible to the naked eye could have been responsible.[18]

Dr. Francis Wrightson concurred that strychnine should be detectable in the blood and the tissues, and that Taylor's implication that it was "used up" in poisoning was incorrect. (Another witness says that he had "never heard the theory until today.")

Mr. Robert Gay described the death the previous year of a bus conductor called Foster from idiopathic tetanus, with convulsions lasting a fortnight: "He had no other hurt or injury to his person of any kind that would account for these symptoms." The man had a large family and was very hard-working: "I call it inflammatory sore throat from cold and exposure to the weather. The symptoms became tetanic in consequence of an extremely nervous and anxious disposition.... I consider the brain had been affected and congestion had taken place."

Mr. Richard Partridge criticised the doctors involved in the post-mortem for not having sufficiently investigated the gritty granules on the spine, which he said were a symptom of arachnitis, or inflammation of the membrane surrounding the spinal column. This disease was rare. Although he had never actually seen a case, such inflammation could produce convulsions and death, although usually this went on for months and he had never heard of a case where the patient had died after a single convulsion. When asked forthrightly by the prosecution whether he thought that Cook's death had been caused by arachnitis, he admitted that he did not.

Dr. William McDonnell set new (low) standards for an expert witness. He expounded his bizarre theory that the case was one of "epilepsy with tetanic complications," caused, possibly, by overexcitement, or possibly by depression resulting from an overreaction to overexcitement. This could also be the cause of syphilitic sores, or possibly the result of syphilitic sores; he does not seem too clear.

Q: Do you mean to stand there, as a serious man of science, and tell me that?

A: Yes, the results of sensual excitement; chancre in one of them, and syphilitic sore throat.

Q: Do you ever hear or know of such a thing as chancre or any other form of syphilis producing epilepsy?

A: Not epilepsy, but tetanus. You are forgetting the tetanic complications. (Roars of laughter in court. One of the judges remarks that he heard someone in court applauding. When a man is being tried for his life such a display is most indecent, he says.)

Dr. B.W. Richardson made the equally preposterous assertion that Cook died of angina despite the absence of any signs of heart disease. He described the case of a 10-year-old girl whom he had attended, who died of symptoms very similar to those of Cook, which he ascribed to angina. But he goes on to say that if he had known then as much about strychnia as he has since learnt from listening to the trial proceedings, he would have made sure that an analysis for strychnine had been carried out!

This shotgun approach to explaining Cook's symptoms was counterproductive. The aim of the defence was to stir up as much doubt as possible in the minds of the jury. But the jury must have concluded that the defence experts did not know what they were talking about, and that the only possible explanation was that given by the prosecution, namely that Cook was poisoned with strychnine despite the absence of a successful analysis. The attorney general in his closing address was able to pour unmitigated scorn on the defence witnesses: "It is impossible to conceive evidence more dishonest...it is a scandal upon a learned, a distinguished and a liberal profession that men should put forward such speculations as these.... Do not talk to me about excitement, as Mr. Nunneley did, being the occasion of idiopathic tetanus... they were topics discreditable to be put forward by a witness as worthy of the attention of sensible men."[19]

This concluded the medical parade. The final witness for Palmer, and one whose testimony he could well have done without, was Jeremiah Smith, lawyer and seedy member of the Rugeley racing fraternity, who also happened to be Palmer's mother's lover. Smith was called to testify in Palmer's favour about his movements on the night that he was alleged to have got the strychnine from Newton. But much to his discomfort, he found himself being cross-examined on the subject of the Walter Palmer insurance policy and the attempt to insure the ignoble Bate, on which occasion Smith had been appointed local agent for the insurance office. The attorney general in his summing up was able to call him "a most discreditable and unworthy witness:" "Such a spectacle I never saw in my recollection in a court of justice. He calls himself a member of the legal profession. I blush for it to number such a man upon its roll."

After stating that he did not recall having witnessed a proposal on Walter Palmer's life to the Universal Life Assurance Company, he was disconcerted to have the document itself pressed into his hand. He was asked if the signature on the document was his. "It is very like my signature," he said, "but I have a doubt of it." Then after a pause, "I believe it is not my handwriting; I swear it is not. I think it a very good imitation." He was then asked if he had witnessed Palmer's signature to a deed of assignment and received a payment of £5 for it, and replied that he might or he might not. The attorney general then placed in his hand a cheque for £5 made out to him and signed by Palmer. Smith stated that he might indeed have received £5 at the bank, "but upon my honour I do not know what for." (Laughter). In the words of another who was present:

> The exhibition made by this fellow in the box was disgusting...the witness's attempts to gain time, and his distress as the various answers were extorted from him by degrees, may be faintly traced in the report. His face was covered with sweat, and the papers put into his hands shook and rattled.

The effect of this was to validate Newton's evidence, despite the discrepancies concerning the timing of events on November 19th. One highly probable explanation for Smith's extreme discomfiture, which would shine a better light on him, would be if the signatures were in fact Palmer forgeries, with Smith trying neither to claim ownership, which would incriminate himself, nor to disown it, which would have incriminated Palmer.

The closing address by the attorney general was damning. He was able to ridicule the defence arguments, and thus avoid having to spend too much time on the difficult topic of the missing strychnine. He faced this difficulty foursquare, but put an effective spin on it:

I have no positive proof on the one hand, but on the other hand, my learned friend is in the same predicament—he cannot say that he has negative proof conclusive of the fact of this death not having taken place by strychnia.

The judge's summing-up, although observing the forms of impartiality, was powerfully against Palmer. Campbell was particularly scathing about the defence's scientific witnesses:

I must say that there were examined on the part of the prisoner a number of gentlemen of high honour and solid integrity and proved scientific knowledge, who came here only to speak the truth and assist in the administration of justice. You may be of opinion that others came whose object was to procure an acquittal of the prisoner.... [Nunneley] certainly seemed to me to give his evidence in a manner not quite becoming a witness in a court of justice...

The jury took 78 minutes to return the guilty verdict, which Campbell said he and the other judges found "altogether satisfactory," without bothering to consult them.[20] Palmer was transported in irons to Stafford by the night mail train with a heavy escort of warders. During his three-week incarceration, he became an obsessive object of interest across the nation. It was widely rumoured that he had strychnine hidden inside his ear in order to commit suicide should his solicitor's appeals to the home secretary prove unsuccessful. Pirate prints and medallions began to circulate; one unscrupulous printmaker reactivated a block depicting William Cobbett and obliterated his name, substituting Palmer's.

When the governor of Stafford prison asked him whether he was guilty, Palmer replied that he did not poison Cook by strychnine. The governor said this was no time to quibble, and asked him again if he had poisoned Cook at all. Palmer repeated that Lord Campbell had summed up for death by strychnine and that he denied the justice of the sentence. This has served to reinforce posterity's view that the prosecution were right in their submission that Palmer had used both antimony and strychnine, and possibly other poisons including morphine too, and that he was taking refuge in the claim that it was antimony, or something else, that had actually finished Cook off.

Lobbying continued while he was in prison. There was a public meeting to call for a delay in execution of the sentence because of the forensic uncertainties; it ended in the summoning of a policeman amid scenes of "indescribable confusion." A Mr. Baxter Langley said that if Palmer were executed, it would be "to satisfy a scientific hypothesis" (cries of "no, no" and uproar).[21] The home secretary refused to receive a deputation from the meeting.

Palmer was hanged in front of Stafford gaol at eight o'clock on the morning of Saturday June 14th, 1856, at the age of 31. Throughout the night, the streets of the town were filled with the sound of tramping feet heading for the prison, and 350 policemen were on duty, including 200 special constables.[22] Most of the spectators, the *Times* reporter commented, were from the labouring classes, but the considerable number of umbrellas that appeared when it started to rain testified that a large proportion were from the better class of person. There was a hush when Palmer appeared dressed in prison clothing with his light sandy hair cropped short, which with his large head gave him an air of unnatural repulsiveness. When invited to stand on the trapdoor, he was reported to have said, "Do you think it is safe?"

A phrenologist, Mr. Bridges from Liverpool, was allowed to take a cast of the head, then the corpse was buried naked in the prison yard without a coffin. In Cork, where prayers were offered up in churches for Palmer's last-minute redemption, there was rejoicing in the streets when it was rumoured that they had been successful. A misprint in one of the Dublin newspapers had accidentally substituted part of a speech by Napoleon III for Palmer's last words, so that he was reported to have left this world with the valediction, "I know that one of the best means to deserve it is to testify my veneration for the Holy Father, who is representative of Christ on Earth."

We will now never know how many people Palmer actually poisoned. Whilst he was awaiting trial, the family vault at Rugeley parish church was opened and the coffins of Anne and Walter Palmer were carried across the road to a small room in a public house and opened in the presence of no less than 30 persons. Anne Palmer, who had been dead for 15 months, was in a porous oak coffin; the corpse was more or less dried up and the smell on removing the lid was bearable. Walter Palmer had been buried in a lead coffin five months before, however, and when a hole was bored in the lid the stench was so appalling that most of those present began to vomit, and one man was ill for four days. Analysis showed considerable amounts of antimony in Annie's body, but no poison could be detected in Walter's.

There is strong circumstantial evidence in the case of Annie, Walter (probably by prussic acid), and Mrs. Thornton (probably by a soporific poison such as laudanum), who died within a fortnight of coming to stay with the Palmers, leaving them money. Also a Mr. Bladon, whom Palmer owed £800 for racing bets and met his end while staying with Palmer and was rapidly buried by him; and of an illegitimate child whom he had by a woman in Rugeley, and possibly others, including Palmer's uncle. Even as far back as 1846, the inquest at Rugeley into the death of a man called Abley, with whose wife Palmer was sleeping, cast doubts over whether the considerable amount of brandy that Abley had drunk before he died had been wholly the result of a misfiring student prank, as Palmer had insisted.

Not only this, but the Palmers had five children, of whom the four youngest died suddenly of convulsions soon after birth. His eldest child William survived. Freudian speculation about Palmer projecting onto his family his own desire to have been an only child and thus monopolising his dissolute mother may be unfounded, for while she was still alive Annie was fearful for the life of young Willie as well as for herself. The truth seems to be more prosaic. Palmer killed his children because he did not want to support them. They were very late pregnancy terminations.[23]

The judicial proceedings were highly unsatisfactory in many respects. Palmer declined to speak at the inquests, so his voice was never heard. There is a strong suspicion that in the best traditions of British justice, the judges got the result they knew was the right one without sticking too punctiliously to the rules. As the *Illustrated Times* remarked, "The prisoner has been as completely (although not so fairly) tried as he could have been anywhere in the world." Palmer was right when he thought that the lord chief justice was biased against him. Campbell, a Presbyterian, apparently also heartily disliked Shee, a Roman Catholic.[24]

Elizabeth Mills moved to a job at Dolly's Hotel in London shortly after the murder, and met with Mr. Stevens several times. Cross-examined, she claimed that the impending trial had not been discussed, and that he had only been asking her "how she was getting on in London" and so forth. The defence counsel was rightly incredulous; Stevens was "not in your station"; he was a gentleman.

But the most unsatisfactory evidence was the scientific. The experts on either side of the case just did not know enough chemistry or toxicology even to state clearly what it was that they did not know, and many of them were more interested in making a show in court than in getting to the truth. As Taylor later remarked, "It is not at all improbable...that had the chemists for the defence changed place with the chemists for the prosecution, the prisoner would have been chemically convicted by his own witnesses. One of the worst effects produced...was the impression left in the public mind that with a little search *medical men might be got to prove anything.*"[25]

It would take many years of patient unravelling to untie the knot of strychnine.

Endnotes

1. Forbes, 1985; Burney, 2000; Russell, 1983.
2. Knott, 1912.
3. *Punch*, 19 April 1856:157.
4. Christison relates how as a student in Paris he heard a lecture on tetanus by the famous physician Dupuytren, who was scathing about those who thought that the disease was caused by inflammation of the spinal nerves. There was nothing to be seen on dissection. How, Dupuytren argued, could such a savage disease arise from an invisible inflammation? The disease was certainly a mystery (Christison, 1885).
5. Udwadia, 1996: 624.
6. Coley, 1991.
7. After the death of Nathaniel Button in Essex in 1848 during an epidemic of arsenical poisoning, Taylor had failed to find any in the body, to the consternation of the authorities (the *Times*, Oct 4 1848:7). Later, in the trial of Dr. Smethurst in 1859, he made a mess of the forensic testing for arsenic. Smethurst, who was almost certainly guilty, was convicted but reprieved by the Home Secretary in view of the medical doubts. Smethurst very probably outwitted Taylor; he knew that perchlorates interfered with the test for arsenic that

Taylor preferred, and administered them to his victim. Then at the inquest on the wife of Dr. Alfred Warder in Brighton in 1866, Taylor expressed his definite opinion that she had been poisoned with aconite, although his forensic tests had failed to detect aconite or any other poison. Warder committed suicide.

8. See also Chapter 8. Otto (loc. cit.) showed that tartar emetic interfered with the detection of strychnine, but only because of its tartrate component. This could be destroyed by treatment with sulphuric acid. But Otto's work was not published until shortly after the Palmer trial.

9. Christison, 1885.

10. Mention of the boar produced laughter in court. The creature provided a link with strychnine's earliest days. Christison (born 1797) studied in Paris, where he knew, or studied under, Vauquelin, Pelletier, Caventou, Orfila, and Magendie. A fighting wild boar of the Combats des Animaux was no longer fit for military service, and was dispatched by Magendie by means of an injection of strychnine into the pleural cavity.

11. This word was presumably a transcription error. It almost certainly refers to the powerful Upas poison of *Strychnos tieute*. The local Javan name for this is Tschettik, according to one source.

12. After writing this I came across a reference to a case in the Indian *Medical Gazette* of 1885 in which a child had been poisoned by a drug peddler substituting nux vomica bark for that of *Holarrhena antidysenterica*, widely used to treat fevers. The author suggests that many of the reported deaths from tetanus in Calcutta at that time may have been due to strychnine.

13. Fletcher, 1925. Campbell, incidentally, served as MP for Stafford, which he called "The dullest and vilest town in all England" (Campbell, 1881).

14. Fletcher, 1925 and also note in the *Times*, 2nd May 1856. Wilkins had got into debt and fled to Dieppe to escape his creditors (Lewis, 2003: 90) He was back in England though to represent William Dove at his trial soon afterwards.

15. According to Newton, the reason was that he had been intimidated by Palmer's brother George, who had told him that the family was taking out proceedings for perjury against a druggist's assistant in Wolverhampton who had said that he had sold prussic acid to Palmer at the time that Walter Palmer had died.

16. These were found at the third post-mortem. One possible explanation would seem to be miliary tuberculosis (tuberculosis in the form of granules containing colonies of slow-growing bacteria). Symptoms are vague and could have accounted for Cook's feelings of general ill-health, but could not have produced tetanic symptoms.

17. Letter from John Vining to the *Times*, 4 June 1856.

18. Letheby's evidence also could be far from reliable. At the trial of Ann Merritt in 1850 for poisoning her husband, a doctor in the courtroom was so astonished by his evidence that he wrote to the Home Secretary. As a result, Letheby revised his opinion, and Merritt was reprieved.

19. The *Pharmaceutical Journal* (1856:16:5–11) said "It will always be observed that if the highest medical authorities are of one opinion the opposite opinion is sure to be maintained by members of the profession who are of no authority whatever. To maintain a thesis against such a man as Sir Benjamin Brodie is a distinction for an obscure practitioner." An American commentator not long afterwards wrote, "It is a singular fact that in recent trials for murder by poisoning, men have been found who, having some knowledge of chemistry, either from ignorance or some other reason, wholly misrepresent the true state of that science. These experts not only hazard, and sometimes even defeat, the ends of justice in the cases in which they are engaged, but their evidence is afterwards, in some cases at least, quoted by lawyers for the purpose of acquitting some notorious criminal. In this manner these false statements may continue to work evil for years after they were first promulgated." The writer was T.G. Wormley, a meticulous worker who appeared as an expert witness in the Freet case and gave the most exhaustive account of the tests ever given in open court (*Ohio Medical and Surgical Journal*, 1864, vol. XVI).

20. The jury sat around in silence for a while until the Foreman gave out pieces of paper and told them to write their verdicts down. They all read "Guilty."

21. *Times*, 11 June 1856:5.

22. *Times*, 16 June 1856:9.

23. A speculation put forward recently by Palmer apologists is that Palmer and Annie might have been rhesus-incompatible. This would have led to any children after the first falling victim to antibodies and dying young.

24. Anticatholicism was intense in Britain during the nineteenth century. Christison relates a case in 1871 when a wealthy man died suddenly, leaving his estate to the Catholic church. Christison found a blood clot the size of an orange in his brain. The magistrate was not satisfied and said that the Jesuits had poisons "unknown to Dr. Christison," who was the leading British forensic scientist of his day. (Christison, 1885).

25. Taylor, 1856.

CHAPTER 12

Mrs. Dove's Brush with the Media

The printing press is either the greatest blessing or the greatest curse of modern times, one sometimes forgets which.

J.M. Barrie, *Sentimental Tommy*, 1896

William Windham in 1797 deplored the fact that newspapers were now avidly read in every ale house. He said that he "never saw anyone of low condition with a newspaper in his hand without comparing him to a man who has swallowed poison in the hope of improving his health."

Half a century later, the *Times* totally dominated the newspaper market. It sold four times as many copies as its three biggest competitors put together. It had no political convictions and published anything that would improve its circulation, raising the major, and probably forever unresolvable, question, of the extent to which it molded mid-Victorian public opinion and the extent to which it merely reflected it.[1] Its power had become "frightening," and it was accused by some of causing the Crimean war.

The Palmer case exploded onto its pages. The printing presses could not keep pace. George Fletcher, Palmer's biographer, recalled having to pay 3s or 4s ($1.00) each day in Birmingham to obtain a copy during the trial; up to twelve times the official cover price of 4d. All of the forensic evidence was dissected in its columns, and much that appeared there was highly prejudicial. The *Times* was not alone; in the year of the trial, 1856, the

167

British Medical Journal[2] had 30 items on the Palmer case, and devoted more than 50 pages to it. This was the year in which the tax on newspapers was abolished, and more than 200 new titles appeared; the inquest on Ann Palmer was packed out with scribblers from at least thirty papers. The *Illustrated Times* published a Special Rugeley Issue. "I suppose you are going to pick us to pieces again in one of your Lunnun papers," a red-faced local demanded of the paper's reporters when they arrived in town.

Even before the trial came to court, the Palmer faction had been massaging the media. An item appeared in a Shrewsbury paper stating that by an "extraordinary coincidence," a gentleman had been seized with sickness "resembling that of the late Mr. Cook," after drinking brandy and water at the Raven Hotel. It was supposed that the liquors being sold there contained something noxious. The paragraph was found to be pure invention, and "on enquiry was traced to a person strongly interested in the defence of Palmer."[3] Then, shortly before the trial, the Birmingham and Staffordshire papers reported that Dr. Letheby in London had investigated a case of tetanus in which strychnine had been suspected, but none had been found. This false report too, according to Taylor, had been placed by the Palmer camp. A pamphlet melodramatically entitled *The Cries of the Condemned, or Proofs of the Unfair Trial and (if executed) Legal Murder of William Palmer &tc.* was printed and circulated over the spurious signature of Thomas Wakley, the coroner and reformer; he said that it was a "vile fabrication."[4]

At the beginning of the eighth day of Palmer's trial Lord Campbell expressed his "earnest hope...that the public journals will continue to abstain from any comments on the merits of the case or any part of the evidence." But it is difficult to see many signs of voluntary press restraint before, during or after the trial. The continuing report of proceedings in the *Times* appeared right next to extensive correspondence about the accuracy of Taylor's forensic evidence. Never before, or probably since, had a newspaper been so charged with a scientific controversy. Taylor's impromptu "theory of perfect absorption" was dissected in its columns, as well as in those of the medical press. Other letters dealt with topics such as squashing the misapprehension that had apparently grown during the course of the trial that antimony would inter-fere with the detection of strychnine.

The instant book was not an invention of the late twentieth century. By the end of 1856, Taylor had already written and had published a 140-page account of the Palmer trial, most of which is a polemic against his critics.[5] The reason he thought this neces-sary was not just to put his scientific viewpoint. It was his inept handling of the media, as they were not yet called, that made the situation far worse and turned an abstruse, though important, difference of scientific opinion into a public-relations disaster.

While the inquests were in progress, Taylor had been visited by Henry Mayhew, brother of the famous reformer Augustus Mayhew, who came with a letter of introduction from Faraday to ask him for an interview to be published in the *Illustrated Times*.[6] This Taylor gave one evening after dinner, and said far more about strychnine, tartar emetic, life insurance and the Palm-ers than he should have said as a professional man about these *sub judice* proceedings, even in 1856 when there were no legal prohibitions. Cross-examined about the interview in court, he showed all the panicky symptoms of someone desperate to avoid the blame for something he now realised was extremely foolish. "I swear solemnly that I did not know he came for information to be published in the *Illustrated Times*. The publication of that article was the most disgraceful thing I ever knew. It was the greatest deception ever practised on a scientific man," he blus-tered under oath. But both Mayhew and a colleague who was present throughout wrote to the *Times* saying not only had it been made clear to Taylor that the interview was to be published subject to his approval of the copy, but also that they had in their possession the proofs of the article showing clearly Taylor's hand-written alterations, all of which were accepted.[7]

Thus Taylor found himself immured in a far greater controversy than he could ever have imagined when he was commissioned by Mr. Stevens to carry out the examination of Cook's viscera. He had already felt obliged to write a letter to *The Lancet* "contra-dicting several misstatements which were made concerning my evidence [at the inquest]." His letter described in graphic detail the condition of the bodies of Ann Palmer and Walter Palmer, and concluded with a hysterical outburst about poisoning: "I have never met with any cases like these suspected cases of poisoning at Rugeley. The mode in which they will affect the person accused is of minor importance compared with their probable influence on society. I have no hesitation in saying that the future security of life in this country will mainly depend on the judge, the jury and the counsel who may have to dispose of the charges of murder which have arisen out of these investigations." At Palmer's trial, Shee unwisely tried to have this concluding paragraph, which was not particularly helpful, read out; but he was compelled by the judges to have the whole letter read, which with its details of the other presumed murders was highly prejudicial to Palmer.

Taylor protests too much. He is understandably sensitive to the charge that he is a less than competent analyst. But he argues both that the stomach he had to work with was so imperfect that strychnine would have been impossible to detect by anyone, *and* that strychnine is destroyed as it carries out its work of poisoning, so that it is no longer detectable. As regards the first point, Taylor may have been unfairly attacked, for there were other cases in which poisons including strychnine were definitely administered, but no forensic traces could be found by nineteenth-century methods, which were still evolving. But Taylor maintains that some poisons are in principle undetectable:

> Is there any chemical process by which the poison of the ordeal bean of Africa, or even of the common laburnum, the seeds of *Ricinus communis*, the poisonous fungi, Darnel, and the sausage-poison of Germany, the poison of the oenanthe crocata can be separated and demonstrated to exist after death in the blood and tissues?

In principle, yes, and he is clouding the issue by conflating the practical limitations of Victorian science with his incorrect ideas on the fate of drugs. The reviewer in the *British Medical Journal* seized on this point: "It appears to us that he sustains his argument in a great measure by adducing what could *not* be done 30 years ago."

The second point is more interesting from a modern viewpoint. He was represented by the defence as having said that in the act of poisoning, the strychnine would have been "all used up." Did he actually say that, and if so, did he stick to that viewpoint afterwards? In the transcript of the trial given in *The Lancet*, which is likely to be the most reliable source, the following exchange occurs:

Q: How do you account for the absence of any indication of strychnia in cases where you know it was administered?

A: It is absorbed into the blood, and is no longer in the stomach. It is in great part changed in the blood. (Liebig, the leading biochemist of the era, thought that the blood was the main organ in which metabolic activity took place.)

Q: Supposing a minimum dose, which will destroy life, has been given, could you find any?

A: No, it is taken up by absorption, and is no longer discoverable in the stomach.

So, yes, he did state in court that the second reason he did not find any strychnine in Cook's stomach was that it had all been consumed in the act of poisoning.

The defence witnesses, on the other hand, argued that strychnine was indestructible, a view that Taylor finds preposterous. "Strychnine, the jury were told, although an organic compound of four elements, carbon, oxygen, nitrogen and hydrogen, was quite indestructible under any circumstances." The defence experts were wrong to some extent; strychnine is not indestructible and undergoes metabolic breakdown in the tissues. But such metabolism is not an inevitable consequence of the fact that it is a drug (poison). Taylor's theory that it is "used up" by the sheer act of poisoning is wrong. He fails to distinguish between being used up in the body in the act of poisoning, and partially disappearing in some other way from the tissues, including excretion unchanged through the kidneys. The following passage from his book shows that he is still wedded to his theory:

To suppose that it alters or affects the blood without in itself undergoing some change, is contrary to the generally admitted doctrines of chemistry. Is it in accordance with the laws of chemistry that A affects B, without B affecting A?

The fallacy is in assuming that the interaction between drug and tissue receptor is a chemical one in this strict sense of the term. Scientists were by now familiar with catalytic reactions. A piece of platinum can make an infinite amount of hydrogen and oxygen combine to make water, without being changed in any way, as discovered by Humphry Davy in 1817.[8] A drug or poison could catalyse a change in the body tissues without itself being changed. As Nunneley pointed out in a cutting letter to the editor of the *British Medical Journal,* many of these facts about drug action and metabolism were already known, albeit poorly understood, in 1856, and were inconsistent with Taylor's theory. In any case, said Nunneley, supposing it were in principle possible to adjust the dose of strychnine to exactly that required to kill the patient and leave nothing over. Does anyone seriously believe Palmer, with a medical training amounting to little more than "strolling the wards" for a few months, capable of doing it? "If Cook died of strychnia, depend upon it, the not finding of it did not depend upon any extraordinary skill in William Palmer, but in the want of skill or proper care in others, whose after thoughts are intended as a cloak to hide primary and fatal faults!" Letheby was equally scathing, and it is difficult to imagine a correspondent in a medical journal today referring to someone else's work as "the most contradictory and absurd thing I ever met with."[9]

The emerging discipline of forensic science had certainly had its analytical failures. In the Castaign case, tests by Orfila, Vauquelin, Magendie, and several others had failed to detect any poison. Taylor cites a number of other cases, and does his own profession a disservice by saying that many more poisonings would be discovered if the forensic scientists paid more attention to determining "in the abstract" whether poisoning had taken place, regardless of analysis. Taylor said, "My opinion was then, and is now, that we may more safely trust to pathology and physiology than to the crude speculations of chemistry."

In that case, why bother with chemistry at all?

So in the Palmer case chemistry in the end proved nothing. At the beginning of the trial, the *British Medical Journal* said, luridly, "If William Palmer is hanged, it will be the test-tube and the retort that will hang him. Chemistry will be the judge and will pronounce the last terrible sentence." But after the trial, it was chastened: "How is it, we ask, that science comes so lamely out of this struggle between life and death?"

The question that the court eventually had to decide was, can pathological symptoms be taken as proof of poisoning in the absence of a positive analysis? Cook's symptoms were fully consistent with strychnine poisoning, but the question also had to be answered as to whether they might also be consistent with something else. The reviewer in the *British Medical Journal* criticised Taylor for not arguing the case step by step: "Instead of pooh-poohing this evidence (about tetanus), Dr. Taylor should have demonstrated its inapplicability to this case." In the end, Taylor had to fall back on the overwhelming corroborative evidence that Palmer was, indeed, a very wicked man. Had he not once owned a horse called "Strychnine"?[10]

A pivotal figure in the tragedy that unfolded in Yorkshire during the days of Palmer's notoriety was one Henry Harrison, self-proclaimed Witchman of Leeds, water diviner, caster of nativities, believer in the stars and dentist.[11] But the person who paid the price for his stupidity was Harriet, wife of a certain William Dove. A comment by the judge at Dove's subsequent trial indicates what the bench thought of this:

> *Harrison:* Sometimes, Dove would say to me, the devils are comfortable now, but they are still in the house; is there no way of getting rid of them? I did not encourage him to think I could rule devils. That was his own fantasy. I thought I could cast his nativity. I believe in the stars.
>
> *His Lordship:* You do. And you find a great many others do the same thing, I dare say.

Dove was not wholly unintelligent; he could read and write well, but he had several screws loose. From an early age he had been an enthusiastic and ill-natured arsonist and animal torturer, described by one schoolmaster as being "of very low intellect, great want of moral power, and of evil and vicious propensities," a tendency which was tackled in the time-honoured manner of flogging him repeatedly. "I thought him insane," the man opined, "but did not feel myself in a position to object to his being flogged," the Victorian answer to attention deficit disorder. Another said, "He had a glimmering of intellect, but was as near an idiot as could be." When Dove's father set him up as a farmer, he was said to have reaped his corn when it was still green, to have lain on the floor and cried without a cause, and to have terrorised his farmhands and neighbours by pointing loaded shotguns at them. Eventually, after selling his farms and moving several times for no sensible reason, he gave up farming and moved to a house in Leeds where he became a drunk. Despite all this he found someone willing to marry him, but before long he and Harriet[12] Dove were living apart under the same roof and on the point of splitting up; unfortunately for her, a lawyer dissuaded her from pursuing a legal separation. Dove's aberrant behavior did not abate, although some of it always had the flavour of silly jokes; he painted the cats with phosphorus so that they glowed in the dark, and locked them in the parlour to frighten the women. But when his mother-in-law came to stay, he held a pistol to her head then fired it out of the window, although she denied rumours that he had once thrown a cat up in the air and shot it.

The role of the imbecilic Harrison in what happened next did not fully come out at Dove's trial, but is described in two written confessions left by Dove. About August 1855, he says, he had some unpleasantness with his wife and went to see Harrison, who told him that he would ensure by means of his magic powers that everything would turn out right. Nothing was better after a couple of days. Dove wrote to Harrison telling him so, but Harriet intercepted the letter and substituted a sheet of blank paper. The next time Dove saw Harrison, he explained what had happened, and Harrison told him that he would never be happy until his wife was out of the way. He drew Dove's horoscope and explained to him that until the age of 32 everything would go against him, but by that age he would "lose" his wife, marry again and come into wealth. On a subsequent visit, Harrison said that she would die before the end of February or March, he wasn't sure which. Dove pointed out that Harrison had previously said that she would die when he was 32. Harrison said, *before* he was 32; he hadn't said how much before.

At about the turn of the year, Dove went to the New Cross Inn in Leeds and Harrison came in with a copy of the *Times* giving an account of the Cook inquest, reading from it to the assembled topers. This was at the time of the acrimonious controversy concerning Taylor's evidence, and whether strychnine would be detectable after death. Dove asked Harrison whether strychnia could be detected. No, he replied, and neither could other vegetable poisons, such as belladonna and digitalis. Dove asked Harrison to get him, or make him[13] some strychnine, as he was troubled with some cats he wanted to get rid of, but Harrison refused. Dove said he would get some elsewhere. In January they met again. Dove told Harrison about his wife's temper and the fact that she was now ill, and Harrison told him not to worry because she would never get better.

Harrison was clearly a good prophet because when Dove met him again on March 6th, his wife had indeed died. There was one small fly in the ointment, however. There was to be an inquest. Harrison asked why, and Dove told him that his wife died very suddenly, and as it was known that Dove had strychnine in the house, she might have got accidentally poisoned. Dove complained, aggrieved, that Harrison had told him a grain or a grain and a half of strychnine could not be detected, but one of Mr. Morley's students had shown him in the *Materia Medica* that it could. On the other hand, Professor Taylor down in London was saying that it could not. What did Harrison have to say about that?

Harrison, slowly waking up, said, "What, have you poisoned your wife?" and Dove replied, "No, I should be very sorry to." Later, when the inquest had opened, at Fleischmann's hotel, Leeds, and several witnesses had given evidence against him, he asked Harrison how the case would go. Harrison said it would be difficult, but he would get him off. For once, his prophetic powers deserted him.

For Dove had poisoned Harriet intermittently with strychnine over a period of about two weeks, until she died on March 1st. The physician, Mr. Morley, said that he thought at first that she was suffering from hysteria, since she was "at the period when women are prone to such attacks;" she had had previous episodes with jerking of the arms and noises in the stomach. But, as he pointed out at the trial, one of the features distinguishing hysterical attacks from strychnine poisoning was that the former were never fatal...

If his confessions are to be believed, Dove was never more than a half-hearted assassin. He got strychnine on at least two occasions from Mr. Morley's pupils and used it to poison animals, promising to one student the skin of a large grey cat to make a tobacco pouch. Soon he was leaving strychnine all round the house, and carrying it around in a razor-case. On another evening, he took some without permission from the jar in Morley's surgery, which was not locked. When the coachman caught him in the act, he turned the gas down and said he had come in to light his pipe, and that Harriet had been suffering from fits, which he mimed by "grinning and throwing up his hands." From February 25th onwards, he was giving her sublethal doses in her medicine and seemed confused as to whether he wanted her to die, sometimes weeping at her side as she lay in bed. He says in his confession that he did not think of the consequences, but when he saw her suffering, it flashed across his mind that he had given it to her, and that she would die. He said, "I cannot disguise the anguish I felt when I returned from Mr. Morley's and found her dead." Palmer's case first called his attention to strychnia, he says, but he should never have thought of using it to kill his wife but for Harrison.

The trial evidence about Dove's guilt was more or less open-and-shut. Animal poisoning was the main alibi. The dates clearly indicated that Dove used the first ten grains to poison animals; therefore, there was no reason why the second lot of five grains was not used for the same purpose. But this time, the post-mortem by Morley was conclusive and the forensic evidence of strychnine in the stomach, by Morley and Herepath, brooked no argument. The defence plea rested almost entirely on the question of Dove's sanity. Because of a recent example, Counsel says in his opening remarks, there prevails a sort of public panic about strychnine and "the instinct of imitation is overwhelming with the insane."

Was Dove mad? Here we are pitched into a major controversy that ran throughout the whole of the nineteenth century and on until the death penalty was abolished in the United Kingdom in 1965. Cases such as Dove's were major planks in the campaign for abolition: "Abolish capital punishment, and the dispute between lawyers and doctors ceases to be of practical importance."[14] Few expert witnesses were called on the subject of sanity in English trials before about 1830, but by the time of the Dove trial, such witnesses were appearing in over half of the Old Bailey trials in which mental derangement played a part in the defence.[14]

The ground rules for the legal definition of insanity had been laid down at the time of the M'Naghten case thirteen years previously. M'Naghten, who had killed Sir Robert Peel's secretary in mistake for the Prime Minister himself, had been acquitted on grounds of insanity in a controversial decision. The House of Lords debated the case and submitted a series of questions to the judges; their responses still form the basis of English law, despite their nowadays being invoked only rarely.[15] The core of the tests is set out in the answers to questions 2 and 3:

> The jurors ought to be told in all cases that every man is presumed to be sane, and to possess a sufficient degree of reason to be responsible for his crimes, until the contrary be proved to their satisfaction; and that to establish a defence on the ground of insanity, it must be clearly proved that, at the time of the committing of the act, the party accused was labouring under such a defect of reason, from disease of the mind, as not to know the nature and quality of the act he was doing, or, if he did know it, that he did not know he was doing what was wrong.

These rules are highly restrictive and it was rare for the defence of insanity to succeed. Courtrooms often became a battleground.[14] Medical witnesses satisfied in their own minds that the defendant was insane would see him hanged because, in the eyes of the law, "he knew the nature and quality of the act he was doing, and he knew that what he was doing was wrong." In the nine years up to 1893, for example, 256 murderers were sentenced to death in England, of whom only eight were declared insane; a much larger number (102) had their sentences commuted to imprisonment.

The problem for the defence in the Dove case was that they had to prove the proposition that his actions in poisoning someone over an extended period were the result of an insane urge. Three medical witnesses with lengthy experience of treating lunatics gave evidence. One of them said that he thought Dove had allowed his mind to dwell on thoughts of his wife's death for so long that he had developed, not an irresistible impulse, but an irresistible *propensity*, to kill her. This was a difficult one to get past the M'Naghten rules. How can you distinguish an irresistible propensity to kill someone from just wanting to kill them?

A later writer on the case attacked the defence argument like this. Suppose Mrs. Dove had suffered a serious railway accident, was lying in bed mortally injured, and everyone knew that she was going to die. Would Dove *still* have had an irresistible urge to give her strychnine?

The arguments for insanity did not cut much ice with the judge, who summed up against the prisoner. Delusions in themselves were not enough. Suppose a man to be labouring under the impression that part of his body was made of glass and he robs someone. Why should he escape punishment because he was irrational in other respects? Why punish crime at all? It was to hold out an example to others. If you punish an insane man, you hold out no example because he thought that what he was doing was right. But a man with weak mind and strong animal propensities would not be deterred if he escaped punishment, or saw others of similar inclination getting off.

The *Times* in an editorial was even more cutting:

Mrs. Witham goes in and finds Mrs. Dove with her body arched and her limbs rigid, exhibiting all the symptoms with which the public has now such a terrible familiarity. The history of his [Dove's] life is ransacked for evidence, and we have schoolmasters and farmers coming forward to prove— what we might easily have guessed—that he was always brutal, mischievous and malignant! Are we then to discuss the intellectual qualities of murderers and hang only those distinguished for diligence and ability?

This evidence that Dove was brutal, mischievous and malignant, the leader writer goes on to say, is exactly what would be brought up in other countries by the prosecution to show that Dove was guilty, being just the kind of man who would be likely to poison his wife!

Even at the time of the Dove trial, liberal opinion was stating a contrary view. The position of the alienists (psychiatrists) was well expressed by the famous barrister Marshall Hall more than sixty years later. In the trial of Frederick Holt in 1919, he asked the court to rule that Holt had killed as a result of uncontrollable impulse. "Will is different from reason," he submitted. A man may know the difference between right and wrong and appreciate the nature and quality of his act and its consequences, and yet be deprived of that instinctive choice between right and wrong which is characteristic of a sane person. Under the M'Naghten rules, defective reason, intellectual insanity, was still the only kind recognised by the law; but Hall submitted that a man's reason, even his judgment, may be clear, yet his willpower may be absent or impaired or suspended. (Holt had been treated for syphilis and shell-shock.) As one angry alienist had said soon after the Dove trial, "As there is a true and a false religion, so there is a medical psychology and a legal psychology.[16]

The jury in the Dove case returned a verdict of guilty, with a recommendation to mercy on the ground of his defective intellect, but Dove was hanged anyway. While in prison, he wrote his confession and talked quite sensibly about his feelings, but also wrote a letter in his own blood, which was found in his pocket:

Dear Devil:

If you will get me clear at the assizes, and let me have the enjoyment of life, wealth, tobacco, more food and better, and my wishes granted till I am sixty, come to me tonight.

I remain, your faithful subject.

William Dove

The relationship between the Dove case and the Palmer case was a symbiotic one, and surely unique. Dove got the idea of using strychnine from the early press reports of the Palmer poisonings, via Harrison, and was able to carry out his crime in time for the evidence about Mrs. Dove's symptoms to be given at Palmer's full trial a couple of months later. Dove's own trial followed on two months after that. Furthermore, the key Palmer witness Elizabeth Mills elaborated her testimony about the effects of strychnine on Cook between the coroner's inquest and the full trial. It was suggested in cross-examination that this was as a result of what she had read in the newspapers about the Dove case.

The *Times* deplored the goings-on in Rugeley, implied that this was evidence of something nasty in the collective woodsheds of middle England, but totally ignored its own role in fermenting the "public panic about strychnine" to which Dove's counsel alluded.

Endnotes

1. Cranfield, 1978; Woods and Bishop, 1983.
2. Known as the *Association Medical Journal* until the end of 1856.
3. Taylor, 1856.
4. Letter to the *Times*, 9 June 1856:9.
5. Taylor 1856.
6. *Illustrated Times*, 27 May 1856.
7. *Times*, Oct 4 1848:7.
8. In the 1820s, before the invention of matches, Döbereiner had invented a "Chemical tinderbox" consisting of a pocket apparatus for generating hydrogen from zinc and sulphuric acid and a platinum coil which would ignite the gas catalytically. Over a million had been sold (Thomas and Thomas, 1997).
9. *Lancet*, Nov 1857.
10. A bay colt by Magpie out of Surprise. It raced in 1849, although the owner at that time was recorded as P. Bateman (information from Weatherbys). The 1856 Cesarewich race was won by a horse formerly the property of the late Palmer, with another formerly owned by the late Cooke coming second (*Punch*, 25 Oct. 1856).
11. The best account is given in the *British Medical Journal*, 1856: 222,242,637.
12. Her name sometimes given as Emma.
13. In this, he was almost exactly a century premature. See Chapter 15.
14. Eigen, J. in Clark and Crawford (1994); Smith, 1981.
15. See for example Smith and Hogan (1998):185.
16. Davey, J.G. Journal of Mental Science, 1858, 5:87.

CHAPTER 13

That Clever Dr. Letheby, So Ugly and Terrific

Fie on these dealers in poison, say I; can they not keep to the old honest way of cutting throats, without introducing such abominable innovations from Italy?

De Quincey, *On Murder As One Of The Fine Arts*, 1827

We all harbour secret fantasies about getting rid of other people, quietly and with the minimum of fuss if possible. In the comedy *How To Murder Your Wife*, Jack Lemmon, defending himself in court, draws a chalk button on the courtroom rail and invites the all-male jury to consider what life would be like if they could make their own wife vanish by pressing it. They realise the possibilities and the acquitted hero is borne out of the courtroom in triumph on their shoulders. The difference between the poisoner and the fantasist is essentially one of a moral sense of affect, the appreciation of the value and rights of others, however disagreeable they may be. But an appreciable minority of citizens would press the button if they knew they would get away with it. Part of the mythology of the Victorian age stems from the ease with which people could get hold of poisons, and the hit-or-miss quality of many of the forensic investigations, which made success a distinct possibility.

In Britain, and in other countries too, we refer to the "Victorian Age." This accidental spread of one person's reign accustoms us to thinking of the bulk of the century as of a piece, and we are tempted to accept the Victorians' view of themselves as presiding over an era of evolution and progress with occasional setbacks. But the century was much more turbulent domestically than they would have us believe.

The later decades of so-called "high Victorian culture" have a very different flavour from the first half of the century. Before about the 1850s, the lower classes were left to get on with their lives more or less unhindered, and if that meant the occasional outbreak of poisoning, then so be it. It was more or less what the establishment expected. Miscreants were rounded up and hanged, but this could safely be left to the officers in the provinces. The protection of property and capital was the main concern; until 1830, forgery was a hanging offence, and certainly hunted down more assiduously than the murder of a fellow proletarian.

The Palmer case was one of the pivotal events of the century's social history. The rapidly increasing power of the media gave it far more publicity than any earlier crime. It was brought from the obscurity of Staffordshire to claim the attention of the capital. But most importantly, the perpetrator was a member of the new bourgeoisie, who were making a bid for some of the commanding heights in society: medicine, science and the law. The man was a parvenu, and, to the horror of the establishment, an immensely wealthy one whose family fortune had itself been founded on the kind of rapacity not seen since their own ancestors had arrived with William the Conqueror, and accumulated by robbing a member of the aristocracy, to boot. Who might be next to be poisoned by such a man? Rumours began to circulate that Palmer had also poisoned Lord George Bentinck, son of the Duke of Portland, leading politician and president of the Jockey Club; although a dispassionate reading of the circumstances of Bentinck's sudden death while out for a walk in 1848 makes it clear that Bentinck, a workaholic, must have died from a massive heart attack or cerebral haemorrhage.[1]

Something needed to be done. On July 10, 1856, less than a month after Palmer's hanging, Lord Campbell, the trial judge, was leaping to his feet at the commencement of the day's sitting of the House of Lords to raise with his noble and learned friend on the woolsack a question of great importance. He would not now revert to the facts which had been disclosed during the trial of a recent case, he said, as they must be well known to their Lordships; but he was shocked to say that for too many years past the crime of poisoning had become most alarmingly common.[2]

Campbell went on to describe how poisoning was encouraged by the existence of burial societies, and by the acceptance by life insurance companies of policies written on the lives of people unrelated to the insurer. He said that until recently there had been no restrictions on the sale of any poison, and as a result, poisoning by arsenic had become alarmingly common. An act had been introduced to regulate the sale of arsenic and it had gone somewhat out of fashion; but unfortunately, another poison equally deadly in its effects, nux vomica, had taken its place. A person might go to any druggist's shop in England and buy a pennyworth of it, and had only to say that he wished to poison rats. True, nux vomica was not so powerful a poison as the alkaloid of strychnia, which was extracted from it, but its administration was attended by the same results. In fact, however, they could not only buy nux vomica but strychnine itself from the druggists, without any difficulty. Some restraints should be placed on the sale of these new poisons, and he should be happy to assist the government in framing legislation.

The Lord Chancellor said that the reason the previous act had been restricted to arsenic was that poisons were difficult to define. Furthermore, it was the opinion of one of the most eminent men in the Metropolis (*doubtless Brodie*) that great evil might be done by furnishing the public with a list of seventeen other articles, all of which were quite as deadly poisons as arsenic; but the home secretary had authorised him to state that the question should have his full and earnest attention.

The wheels of the legislative machine were in motion. Successive rounds of lawmaking would secure the safety of the British public and ensure that strychnine became virtually unobtainable, in only a little over a century.

In nineteenth-century Britain, nux vomica and strychnine were widely available. Nux vomica cost eightpence (7¢) an ounce at the pharmacy in 1859. A druggist reported on the occasion of a suicide in 1857 that he had "sold the same (three grains of strychnine) for four or five years to many persons."

Marsden's vermin and insect killer was a salmon-red preparation containing a quarter of a grain of strychnine per packet, mixed with flour and a vegetable dye. Battle's vermin killer was sold in small packets each consisting of an inner and an outer envelope. It contained the equivalent of no less than three grains of strychnine per packet, mixed with flour and Prussian blue. In British homes, it was usually spread onto bread and butter and left on the floor for rats and mice to eat. The mice were often found standing up dead, stiff and erect.

Despite this, however, strychnine was never a particularly popular murder implement, at least as far as detected cases go. A fairly reliable record can be pieced together for the decades after 1856, for by now the medical and legal authorities were more alert to its possibilities. The statistics show that it vanished from the scene from 1856 until 1870, during a period when the Palmer trial was still fresh in the public consciousness. Then there was a burst of activity, with at least eleven criminal cases recorded between 1870 and the end of the century, and for this short period it was the poison of choice. Of the eleven, only one (Thomas Cream) was a doctor with access to medicinal strychnine. The rest were all "ordinary poisoners," if there is such a thing, who reached for the means closest to hand.

Most of these cases speak to us more eloquently of the sadnesses of Victorian society than anything of scientific significance. Elizabeth Pearson of Gainsford near Darlington[3] poisoned her elderly uncle with Battle's vermin killer just to get her hands on his furniture, moving it out of his house and into hers as soon as the body was cold. "What I have got I will stick to," she said, but was hanged for it. Edward Bell, yardman to a farmer at Gedney near Spalding[4] was so keen to dispose of his wife with a shillingsworth of strychnine that he left her funeral procession to send a telegram to his paramour Mary Hodson,[5] arranging to meet her that night so that their engagement could be announced. The jury was out for less than two minutes, after hearing from Dr. Stevenson that the body contained enough strychnine to kill three people. Mary-Ann Britland,[6] an Ashton-under-Lyne factory worker, vehemently denied poisoning Mary Dixon, with whose husband she was having an affair, but failed to explain why she had asked everyone, with great agitation, can they tell if she has been poisoned? Can they tell it in mouse powder? Can they tell it in tea? William Waterhouse testified that he went with the prisoner to the chemist to buy three twopenny packets of mouse powder, and said jokingly, "Now, Missus, administer them scientifically, in not too large doses, and then you'll secure the club money," to which she replied, "Nay, Master, if you thought I would do anything of the sort, may God forgive you." She got £11/17s on her husband's life from the Prudential Assurance, and £8 from the Society of Odd Fellows. John Butterworth, a hatter, said that shortly after the murders, she and Thomas Dixon came into his shop and asked to change a hat he had recently sold to Thomas Britland for one that would fit Dixon, which he agreed to do for five shillings. Mary-Ann spent the period before her execution singing hymns, wringing her hands, and protesting her innocence of the crime of killing her rival. But she was silent on the subject of whether she had also poisoned her husband and her daughter; for a feature of the trial was the amount of evidence that was admitted about these two other deaths, with which she had not been formally charged.

One of the most poignant cases was that of George Horton, a miner.[7] A widower aged 37, he and his family shared two cottages knocked into one, at Swanwick near Alfreton in Derbyshire, with the Bowskill family. Three of his children slept in one bed in an outer bedroom, and he slept with his eight-year-old daughter, Kate, and a two-year-old, in an inner room. After he left early for work one Monday morning in 1889, Kate went into the outer room and got into bed with the other children, complaining of pains, and said that her father had given her something blue to drink out of a cup. The defence said that no such cup could have been used, for all three that the family owned were downstairs in the kitchen. But Horton never went to work; he turned round and went back home as if he knew what to expect, showing no emotion, but later made a full confession that he had poisoned her for £7 insurance, and was hanged at Derby.

Not all paid the ultimate penalty. Henry Bowles was reprieved because there was uncertainty about the justice of his sentence, and his guilt was never conclusively proved. The extraordinary

Christiana Edmunds, on the other hand, escaped the noose because, although she was clearly guilty, she was at the very least on the cusp of lunacy, and the criminal system really had no idea what to do with her.

Bowles, who was 53 years old, was head gardener to Dr. Müller at Camberley in Surrey. His household seems to have been a peculiar one; there was his wife, whom he referred to as his housekeeper, and his son, whom he referred to as his nephew. On September 22, 1887, he called on Dr. Twort to say that his wife was in a fit, and he thought she was dying. The doctor asked why he hadn't sent his son with the message, and Bowles replied that, as a matter of fact, his son wasn't too good either. At the house, Twort found both of them expiring from strychnine poisoning, and a pill box on the mantlepiece labelled "bilious pills," which almost immediately disappeared. According to Bowles, his wife as she lay dying had said to him, mistaking him for his son, "I hope if anything happens to me, your father will not be blamed." Dr. Stevenson, who did the autopsy, found ¼ grain of strychnine in Mrs. B. and ½ grain in the son, but no trace of mercury such as would have been found in bilious pills, and no sign of fungus poisoning (the house was surrounded by mushroom woods).

The motive that the prosecution suggested was that Bowles was attached to a fellow-servant from a former job, and had asked her recently if she was married, saying that he had a comfortable situation living with his son, but the worst of it was an old housekeeper who lived with them. But in an expressive piece of Victoriana, it was pointed out in court that the pair were not legally married, and Bowles could have "Sent her away at any time."

Bowles had once worked in Derbyshire where it was common for landowners to distribute strychnine to their gamekeepers. Bowles or his wife had probably brought it to Surrey. But which of them, and more importantly, who had administered it? Arrested for murder, Bowles said, "I am afraid she has laid hands upon my poor boy," having burnt two letters that were in his pocket, saying they were of no relevance. The house was full of more empty pill boxes that had once contained various poisons, and the scene-of-crime evidence appears to have been a total mess. Bowles was convicted but later reprieved when many supporters produced evidence that the outwardly stable Mrs. Bowles had often been subject to suicidal impulses.[8] The case was apparently never solved. In the words of Bowles's defending counsel, "It is one of those mysteries which human ingenuity and the human mind are unable to unravel, and the truth is known only to the Supreme Being."[9]

If a case involving the death of a four-year-old boy can be said to have its lighter aspects, then the 1871 Brighton poisoning scandal fits the bill.[10] Christiana Edmunds, a 34-year-old spinster, was tried at the Old Bailey under the Palmer Act for the murder of Sidney Albert Baxter. The courtroom was crowded, and many persons of consideration, including ladies, filled the reserved seats. The reporter described Edmunds as of ladylike appearance and becoming manner.[11] She pleaded not guilty and anxiously studied the faces of the (all male) jury as they were sworn in.

The prosecuting counsel began by saying that the accused lived with her mother in comfortable circumstances. Some time earlier, she had formed an acquaintance with the family doctor, Dr. Beard, which "seemed to have ripened into a state of affairs of a kind that did not ordinarily exist between a medical advisor and his patient."[12] In September 1870, after the attachment appeared to have subsided, the prisoner allegedly gave Mrs. Beard a chocolate cream of very disagreeable taste and nauseous properties. After considerable suffering the lady recovered, but suspicion had fallen on Edmunds. It was the prosecution's case that everything else that had occurred in Brighton was the result of actions by her to divert suspicion from this attempt to poison Mrs. Beard, which had launched her on a course of action "probably unparalleled in any court of justice."

Charles Miller said that he had been staying with his in-laws, the Baxters, at Brighton on June 12th the previous year when he went to Maynard's sweet shop in West Street and bought a shillingsworth of chocolate creams. His nephew Sidney was given one and died twenty minutes later. Some idea of the Victorian acceptance of food adulteration can be gained from the fact that the boy and other members of the family had already eaten the sweets earlier. They suffered all of the symptoms with which we are now so familiar: dizziness, stiffness of the joints and nausea, yet they took no further action, and still gave one to the unfortunate child.[13] Letheby, whom Christiana in her sole surviving letter to Dr. Beard described as "that Clever Dr. Letheby, looking so ugly and terrific," gave the forensic evidence. "Good God! This is filled with strychnine!" he is said to have exclaimed on examining one of the chocolates.

A Brighton chemist, Isaac Garrett, said that the prisoner, whom he did not know by name, often frequented his shop, and one day asked for strychnine for cats. Despite his initial refusal, he eventually gave in and sold her no less than thirty grains on three occasions. The 1868 Sale of Poisons Act, which had only come into force three years before, stipulated that no one should be sold anything on the poisons list unless either personally known to the pharmacist, or introduced to him by someone he did know. The legislators had reckoned without the ingenious Christiana, who produced as witness to her false signature a shop assistant from down the street, who didn't really know her from Adam.

Garrett subsequently received three visits from a boy, later identified as Adam May, aged 11. On the first two visits, the boy produced letters purporting to be from another firm of druggists, asking him to lend them some strychnine until their own supplies arrived. Again, Garrett was initially reluctant, but was persuaded by the cunning second letter and gave the boy a drachm of it in a bottle to take away. During the inquest on Sidney Baxter, the boy returned with a letter claiming to come from the coroner, asking for the loan of his poisons book; after it was returned, Garrett found that the page carrying Christiana's signature as "Mrs. Wood of Kingston, Surrey" had been torn out. The letters were all produced and certified to be in the prisoner's handwriting.

Adam May gave evidence and said that not only had he run the errands to the chemist for Edmunds, but had also bought chocolate creams for her, and returned them when she had said they were not the right kind. On each occasion she had met him in the street, and once she had paid him fourpence halfpenny for his pains. Other children told similar stories of running errands to and from Maynard's shop. Brighton shop assistants testified to her leaving bags of sweets behind when she went into their establishments, and of becoming sick when they ate them. Another boy, Benjamin Caulthrop, ate twelve chocolate creams that she gave him, and was ill for a month.

Not content with the actual poisonings, she conducted a wonderfully baroque campaign of positive disinformation, which would have engendered admiration had it not been so cruel to Mr. Maynard, and especially to poor Sidney's family. She went into the sweetshop and caused scenes. At the inquest on Sidney she gave detailed and highly coloured accounts of how Maynard's products had poisoned her friends, and stridently demanded that something be done. Afterwards, she wrote three lengthy letters to Sidney's father, signing them as various anonymous concerned citizens and protesting at the inaction of the authorities. "Such a deadly poison as strychnine ought hardly to be in existence," she said in one of them, echoing the thoughts of many before and since. "Had I lost my child in such a sad way, as a parent I should feel myself in duty bound to take proceedings against the seller of the sweets." When a policeman eventually went to interview her, he found her lying on a couch claiming to have just been poisoned by a parcel sent through the post containing strawberries, a pair of gloves, and poisoned apricots. "I poison Mrs. Beard?" she expostulated. "Who can say so? I've been nearly poisoned myself!" Several other Brighton citizens were sent parcels containing poisoned foods, including Garrett. The pharmacist complained petulantly in a letter to the *Times* that not only had he narrowly escaped death by failing to eat two peaches poisoned with his *own* strychnine, but also had personally been put out of pocket to the tune of £30 by the whole affair. It was not his fault, he argued. "Neither banker nor chymst is proof against cunning fraud."[14] He felt that this powerful poison should be sold to the public "on no pretence whatsoever," since the amount required to destroy a cat or dog without undue suffering was highly dangerous in the public's hands. *The Lancet* was appalled that "this wretched woman" had been able to obtain so much of it so soon after the act that was supposed to prevent just such occurrences.[15]

The prosecution evidence was so strong that Serjeant Parry, defending, could do no more than go through the motions of putting up a defence on the facts. An insanity plea seemed more promising, and to this he switched almost immediately, although he told the jury he was sensible of the inconsistency of trying to argue at the same time that she was innocent, and that she was insane. The defendant's father had been a suicidal maniac dying in his forties, and not only he, but her brother, sister, cousin and grandfather on her mother's side had all suffered from "paralysis of the insane." Christiana's mother gave evidence that she had dreaded her reaching this age, and twelve to fifteen months ago, a great change had come over her; she developed protruding eyes and manic laughter. Even now, said Parry to prove the point, she had the "idiotic vanity to deny her real age," which was 43, not 34. The chaplain of Lewes jail, who had observed her during her incarceration, declared her to subject to entire destruction of the moral sense, having no idea of the seriousness of her situation. But the prosecution insisted on her sanity. Insanity consisted essentially of delusions, they maintained; and it had never been claimed that Edmunds could not know what she was doing, given her cunning.

The judge's summing-up followed the same unbending line as that in the Dove case fifteen years before. There were many diseases to which the mind was subject, he said. There was the idiot, who was born without any mind whatsoever. There was the raging mad person, who had no more criminal responsibility than a tiger. But the most numerous cases were of persons said to be subject to delusions, and he directed the jury that delusions alone were not enough. They returned a guilty verdict.

Earlier in the trial, he had said that Dr. and Mrs. Beard should be entirely left out of the case. He had read their depositions, and there was a good deal of their evidence he should not admit at all. Edmunds was asked if she had anything to say before the sentence of death was passed. Yes, she replied, she wished she had been tried on the other charge that had been brought against her, the improper intimacy said to exist between her and Dr. Beard. It is owing to my having been a patient of his, and the treatment I received, that I have been brought into this dreadful business, she went on. I wish the jury had known the intimacy, his affection to me, and the way I have been treated.

The judge said he was not at all disinclined to believe her statement, and believed that the unhappy circumstances *in which she had placed herself* towards the end of 1870 probably led to the position she was now in, but the truth of that only confirmed the propriety of the verdict. He then passed the death sentence.

The *Times* described the scene in court as one of "romantic ghast-liness" as the prisoner sat demure and seemingly unmoved.

In the hush of the court she was then asked, through a female prison officer, as was required, whether there were any circumstances that might necessitate a delay to the sentence. "Yes," she replied, "I am pregnant." When the wardress told the judge, "She says she is, my lord," there was a sensation. The judge ordered the Sheriff to invoke a mediaeval law requiring the empowerment of a panel of matrons. The undersheriffs swept in with their swords, cocked hats and frills. Twelve women were sworn in and told to determine the question with the assistance of the police surgeon, Dr. Gibson, and Dr. Beresford Ryley of Woolwich, who happened to be in court. The *Times* called the whole procedure a scandal that ought to be, like mediaeval witch trials or trial by battle, assigned to the dustbin of history (which it was, for it was never used again). There was a farcical interlude when the doctors reported that they did not have a stethoscope between them. A police constable was sent out into Fleet Street to buy one, and returned with a naval telescope, saying it was the smallest one he could find.[16]

The jury reported that she was not "quickening" with child; she could hardly have been so, for she had been in custody for several months. "It was not until the last frail straw on which she leant abandoned her to her doom that the full peril of her situation dawned on her for the first time, and the awful aspect of her despair was terrible to behold. The poor ladies who were her unwilling judges wept around her, while she looked from one to the other in mute, unspeakable woe."[17]

Once again, the alienists protested at the sentence and called for a reprieve. *The Pall Mall Gazette* joined the call, but admitted that it had no better reason for desiring this than the fact that she was a woman. The trial judge thought the sentence should stand, but that she should not be hanged until the state of her mind had been gone into, disregarding the inconsistency that it had already been gone into before the jury reached its verdict. Two further eminent doctors examined her, pronounced her of unsound mind, and she was carted off to Broadmoor hospital for the criminally insane, where she died in 1907.

The *Spectator* condemned the reprieve. "The real reason for the pardon was that a weak man thought the papers would scold him if he hanged a woman of sound education and refinement, whose father was mad. "Had Christiana Edmunds been a servant she would have been hanged without more ado." In the opinion of some, then and since, she was not insane at all, but was an unprincipled harridan; the insanity decision was society's way of transforming her from a sinister poisoner into a pathetic, deluded spinster.[18] Seen in Broadmoor more than thirty years later, she was said to remain just as she had always been, taking no interest in her surroundings and engaged in "such childish acts as collecting dust from the bricks of her cell to redden her lips and cheeks.[19]

As this trial took place, the man who later called himself Thomas Neill or Thomas Cream was enrolling as a medical student at McGill University, Canada, but not long after finishing his studies, he arrived in London to study at St. Thomas's Hospital. Twenty years and many miles later, he was to follow Christiana Edmunds into the dock at the Old Bailey.

Late Victorian London was surrounded by a vast ring of seedy older suburbs, such as Lambeth and Waterloo (known as "Whoreterloo"), for which St. Thomas's was the local hospital. In the autumn of 1891, Somerset Maugham was a medical student there, delivering babies in the alleyways and tenements; he must have passed Cream many times.[20]

Thomas Neill Cream was born in Glasgow in 1850, but the family soon emigrated to Canada. Fellow students at McGill remembered him as what would soon become known as a "masher," with ostentatious clothes, flashy jewellery and an ingratiating manner. After graduating, he commenced on an unsavoury career as abortionist in Canada and the United States, interspersed with visits to London where he consorted with prostitutes in the Waterloo area. When he first left Ontario to study in London, he left behind a wife who soon died; he may have poisoned her by mail, sending over poison that she took surreptitiously.

When he returned to North America, he was soon in trouble. A hotel chambermaid was found dead at Cream's premises in Ontario. Cream left for Chicago and was never prosecuted. In Chicago, self-proclaimed "wickedest city in the world," he soon acquired a reputation as an abortionist. Once again he came under suspicion, and was arrested in August 1880 for the murder of Julia Faulkner. But Cream had a good attorney and got off because there was a conflict of evidence between Cream and his illiterate "colored" assistant. Later that year an Ellen Stack died.

He next faced prosecution for sending scurrilous postcards to a man called Martin, accusing him of having fled England on account of siring bastard children, and of passing venereal diseases on to his wife and children. A warrant for Cream's arrest was sworn out. "The average husband and father will be pretty apt to conclude that even hanging would be too good for him should he be proved guilty," the Chicago Tribune opined.[21] Cream skipped bail to Canada, but more serious matters were in the pipeline. He was arrested there and brought back to Chicago to stand trial for murder.

Cream had been advertising a patent medicine for the treatment of epilepsy. One of the respondents was a 61-year-old railway employee, Daniel Stott, from Boone County, Illinois, who unwisely sent his 35-year-old wife Julia to Chicago to obtain his medicine, whereupon she immediately became Cream's mistress. On June 11, 1881 Cream gave her a prescription for her husband, which she took to Buck and Rayner's drugstore. She said that she saw Cream put some white powder into the medicine before giving it back to her. Stott took the medicine and almost immediately died.

No one in Boone County suspected that Stott had been poisoned, and he was buried. Unable to leave well alone, Cream wrote to both the local coroner and the district attorney accusing Buck and Rayner of having negligently killed Stott by making up the prescription with too much strychnine, and demanding an exhumation. Professor Walter S. Haines of Chicago found 3.4 grains of strychnine in the stomach and a further 2.6 grains in the medicine. Cream was tracked down to Belle River, Canada, whence he had jumped bail on the postcards charge, and brought back to Illinois for trial where he was sentenced to life imprisonment.

In June 1891, the governor of Illinois commuted the sentence and Cream was released. The circumstances are mysterious, but his family probably bribed the prison authorities. It was said that anything could be bought in Joliet prison at the time, including drugs or a free pardon. By October 1st he had arrived back in England, and within a fortnight had bought his first strychnine, and poisoned Ellen Donworth, age 19, and Matilda Clover, age 27, both prostitutes, before leaving again for the United States.

No progress was made in the police hunt for Donworth's poisoner, and it was not even known that Clover had been murdered, for she was a penniless alcoholic, incompetently treated by a "club patient" doctor, who certified the death as being due to alcoholism. Further enquiries were only made the following April when suspicion was picked up by house-to-house enquiries after two later murders.

While across the Atlantic Cream did not, as far as is known, poison anyone. But the events of the visit provided some strong circumstantial evidence. At Blanchard's Hotel, Québec, he made the acquaintance of a Mr. J.W. McCulloch who spent several days in his company and learnt a good deal, possibly more than he wanted to know, about Cream's preoccupations. Later that year when he realised that the Dr. Neill on trial in London was the same as Dr. Cream, he contacted the British police.

Cream showed him a small bottle containing a white powder. "That is poison," he asserted, "I give that to women to get them out of the family way." He also showed McCulloch some gelatine capsules and told him that he used them for administering the poison. He showed him some false whiskers that he said he wore when performing abortions, and various pornographic materials. Cream said he suffered from headaches and appeared to be a morphine addict.

Cream bought about 1,000 one-sixteenth grain strychnine pills and a case of various other drugs from the G.F. Harvey Company, who agreed to let him sell its products in England on commission. While in Canada, he also asked a favour of a Montréal shipping agent. Could he collect a small parcel and pass it on to the Purser of the "Labrador" to take to England for him? The man did as requested, but opened it before handing it over. He found it to contain 500 copies of an open letter, signed "W.H. Murray," addressed to the guests of the Metropole Hotel in London, informing them that the murderer of Ellen Donworth was employed by the hotel, and that their lives were in danger if they continued to stay there. The man did not inform the police, assuming that the leaflets were part of an elaborate hoax. This was another of several crack-brained extortion schemes that Cream mounted during the period of the murders. In no case did he follow through effectively to try to get his hands on the money, and the details of these schemes as much as anything else argue for his insanity, possibly owing to the effects of tertiary neurosyphilis.

Early on the morning of April 12, 1892 two more girls, Alice Marsh and Emma Shrivell, died of strychnine poisoning at a seedy address in Stamford Street, Waterloo. Cream sent a letter to the father of his fellow lodger, a student at St. Thomas's, demanding money for not disclosing that his son had been the poisoner. The letter enclosed a press cutting concerning the death of Ellen Donworth, not the two girls in Stamford Street, clearly a clerical error on Cream's part. Eventually, having already booked his passage back to the United States, he was arrested on the charge of attempted blackmail. The police searched his rooms. His medical cabinet contained 54 bottles of pills, which were sent to Dr. Stevenson for analysis. Seven of the fifty-four contained strychnine. Bottle No. 2 was full, containing 168 tiny pills, each weighing three-quarters of a grain. The bottle was marked "strychnine 1/16 grain"; Stevenson found their actual content to be 1/22 grain each. Accompanying the cabinet was a box of gelatine capsules, each of which would easily hold twenty of the pills, a generous lethal dose. Other pills contained sublethal doses of other poisons.

The body of Matilda Clover was exhumed from underneath fourteen other coffins at Tooting Cemetery. Stevenson performed an autopsy, finding strychnine in the stomach, liver, chest fluid and brain, and was able to purify sufficient of it to kill a frog. This was the first time in England that strychnine had been forensically identified in a human victim after death. He calculated that in portions of the organs he had removed, weighing two pounds, there was one-sixteenth of a grain, and that therefore in the whole body, given that Clover had vomited copiously, a fatal dose had been present at the time of death. He did not find any brucine, which meant that the poison had been administered as purified strychnine and not as nux vomica.

Cream's trial for the murder of Matilda Clover took place at the Old Bailey beginning on October 17, 1892 before Mr. Justice

Henry Hawkins, called by one contemporary member of the bar "a wicked judge and a wicked man," and popularly known as "'Anging 'Awkins." Some of the strongest evidence came from a Lou Harvey, whom Cream had attempted to poison in the street after spending the night with her in a hotel. She said that he had given her two capsules to take resembling those found in his effects; she had pretended to take them but concealed them and thrown them away.

The forensic evidence was straightforward. The sole scientific witness was Dr. Stevenson, and the confidence and clarity of his evidence underlines the progress that had been made since Palmer.

Q: Is a vegetable poison usually more difficult to discover than a mineral poison?

A: Yes, it is generally more difficult.

Q: Do you put it that the colour test in strychnine poisoning is uncertain and fallacious?

A: No, but it must be confirmed by other experiments, proof that it is an alkaloid, giving the colour test with several reagents appropriate for strychnine, and the fact that it does produce the physiological test on the frog.

Q: I suppose other substances will present the same appearance as strychnine?

A: Yes. I would not in such small quantities rely on the crystalline form. The form of the crystals does not in this case enable me to pronounce positively; nor does the colour test by itself. The action on the frog shows either strychnine or brucine, I think. ...I take the aggregate of the symptoms; I put the whole thing together, and...come to the conclusion that it was strychnine poisoning.

Q: There have been many cases of strychnine poisoning tried?

A: Yes.

Q: There was the well-known case of William Palmer?

A: Yes. That case was in the infancy of our knowledge of strychnine.

Q: Was there not a great deal of conflict of knowledge about that case?

A: No, I would not say that. We knew little of strychnine then. It was the first homicidal case in which it was used, and some time after I had the first suicidal cases. Of late years we have had many cases. I have been through the Palmer case recently.

Q: May not an inexperienced person make a mistake in the colour test?

A: I think that may be so, if the drugs are not pure. The colour I rely on is a purple-violet, which then passes through a play of colours that I cannot explain. I know nothing else that gives that precise play of colours.

Q: As regards the absence of brucine, so far as it is valuable at all, what would it indicate?

A: It would be an indication that the strychnine had not been administered in the form of nux vomica. I have not experimented in the direction of tracing brucine after death, and I do not know anyone who has.... Having considered the matter carefully in the light of my experience, I have given the result I arrived at.

An expert witness, and there could have been no one in the courtroom who was in doubt that strychnine was the fatal substance.

The jury was out for only ten minutes. A stay of execution was allowed to provide for the arrival from the United States of evidence about Cream's sanity, but when it arrived it was turned down. It is not certain whether he admitted his guilt; one account says that he told the prison staff that he had murdered many other women. The hoisting of the black flag after his execution was greeted with cheering in the streets.

Whereas the case of Jack the Ripper not long before was discussed endlessly at the time and subsequently, there is a very real feeling that Cream, with his pornography and abortions that could not even be mentioned in open court, was just *too* disturbing. There was an element of indecent haste about the dispatch with which he was tried and executed, and a notable lack of debate in the press, including the medical press, on the sanity question. *The Lancet,* unprofessionally for a medical journal, preferred to put the safety of the community first and leave the decision to the home secretary.[22]

Cream was obsessed with strychnine, which he took for its presumed aphrodisiac properties, and which he gave to women to procure abortions. He presumably obtained a sexual thrill from the pain it caused, which must have led to his abandonment of abortions, passing up the career ladder to full-time poisoner. It is curious that he no longer seemed to need to witness the deaths of any of his victims.

His last words on being hanged, according to the hangman anyway, were, "I am Jack the...." He would have made an ideal candidate for the post, but the almost unanswerable objection is that while the Ripper was on the rampage in London, Cream was safely behind bars on the other side of the Atlantic. Ripper enthusiasts are not easily deterred, however. Speculation persists about two techniques that would have enabled Cream to be in two places at once.[23] The first suggestion is that his relatives might have bribed his release for him to be in London to murder the Ripper's victims in 1888, then back in Illinois to be formally let out in 1891. An even more interesting, or preposterous, theory is that there were actually *two* Thomas Creams, using the same identity and cooperating in providing each other alibis.[24] The most convincing argument against him though is that of *modus operandi*. Although Cream was a doctor, he never carved up his victims, and seemed even to avoid staying around to see the corpse.

Endnotes

1. Beaconsfield, 1852; 380.
2. *Hansard*, 3rd series, 143:540; *British Medical Journal*, 1856:615.
3. *Times*, 10 July 1875.
4. *Times*, 5 July 1899.
5. Surely no relation to Palmer's presumed father?
6. *Times*, 23–24 July 1886.
7. *Times*, 30 July 1889.
8. A similar case, apparently involving attempted suicide by the depressed wife of a gamekeeper, took place in Scotland in 1913 (Roughead, 1939: 165–190). The wife recovered and the husband was acquitted of attempted murder. Either he tried to kill her, or, as the jury decided, she tried to kill herself and put the blame on him.
9. *Times*, 14 January 1888.
10. Knelman, 1998: 138–144; 250–251; 256–257.

11. *Times*, 16 January 1872:11; 17 January 1872:12. Newspapers at this period did not carry illustrations, and I have been unable to trace a photograph of her. A sketch made of her during the trial (Figure 14) makes her appear quite normal.
12. *Brighton Herald*, 20 January 1972:3–4, which is the fullest press account of the trial. For much other detail that came out during the committal proceedings, see the *Brighton Times* of Aug. 19, Aug. 26, Sept. 2 and Sept. 9, 1871.
13. Apparently though a doctor was called in the first time the child became ill; he unfortunately reassured the family that the illness was not due to anything he had eaten (Williamson, 1930).
14. *Times*, 31 January 1872.
15. The schedule to the act listed nux vomica seeds and bark, and "The poisonous vegetable alkaloids" as poisons to be regulated. Note the continued vagueness of the term "poisonous." Edmunds appears to have used arsenic and zinc salts to poison the citizens of Brighton with as well as strychnine. One of them, Caroline Walker, gives a pretty clear description of her symptoms of strychnine poisoning, although in court they were described as symptoms of arsenic (*Brighton Times*, Aug. 19 1871; 2).
16. Marjoribanks, 1929.
17. Letter from Dr. J. Beresford Riley to the *Times*, 20 January 1872.
18. The sole surviving letter from her to Dr. Beard, quoted in full at Williamson (1930) and in the *Brighton Times*, appears sane, though neurotic and pretentious, and certainly indicates both that she was strongly infatuated and that he had probably encouraged her. She calls him "My Dear," "Darling," and "Mio Caro," refers to his wife as "La Sposa," and tells him to fancy "a long, long bacio from DOROTHEA." She had obviously been reading *Middlemarch*, the first part of which was published in this same year, 1871. For Beard's vigorous attempts to backpedal from the relationship, see the *Brighton Times*, Aug. 19 1871 p. 2 and Sept. 9 1871 p. 3.
19. Folsom, 1909: 98.
20. Shore, 1923; MacLaren, 1993. A few more details are given by Shore (1935).
21. *Chicago Tribune*, June 19 1881.
22. *Lancet*, editorial, Nov. 19, 1892.
23. http://ripper.wildnet.co.uk/cream.htm.
24. In Marjoribanks, 1930 the biographer relates that the famous barrister Marshall Hall said that early in his career, he was called upon to defend a man accused of bigamy. Several women claimed he had married them in London. Marshall Hall advised the man to plead guilty, but he replied that if the police checked with the authorities in Sydney they would find that he had been in jail there when the alleged bigamies took place. To Hall's astonishment, this was found to be true, and the man was acquitted. Years later, Hall recognised him as Cream. There is no corroborating evidence for this anecdote.

CHAPTER 14

Tigers, Lions, etc.; Six Hundred Kilograms

"Strychnine is a grand tonic, Kemp, to take the flabbiness out of a man."
"It's the devil," said Kemp. "It's the paeleolithic in a bottle."
"I awoke vastly invigorated and rather irritable. You know?"

H.G. Wells, *The Invisible Man*, 1897

At the end of August 1904, thirty-one runners assembled in the newly built St. Louis athletics stadium to run the marathon in the third olympiad of the modern era.[1] The games were inauspicious. There had been an acrimonious dispute between Chicago and St. Louis about which city should host them. On the insistence of President Theodore Roosevelt, St. Louis won, but the local organisation proved to be disastrously incompetent. No foreign country sent a proper team, and the games were virtually ignored abroad. The London *Times* devoted far less space to the whole of the games than to the Dartmouth yacht regatta being held at the same time. A man with a wooden leg won four gymnastics medals.

The arrangements for the marathon seem to us hopelessly amateurish. The eventual winner's time of three hours, 28 minutes and 53 seconds was very slow by modern standards, especially given that the course was only 40 km (24 miles and 1500 yards) long; it was not standardised at its present distance until the next olympiad. But the 1904 race was gruelling, for it went up hill and down dale on dirt roads in the stifling heat of a Missouri summer afternoon. After five laps around the stadium, the runners disappeared cross-country in a cloud of dust, pursued by a mass of judges, physicians and journalists in automobiles, one of which plunged off a 30-foot embankment, causing serious injuries to two doctors.

The eventual winner was Thomas J. Hicks, born in Birmingham, England but now of Cambridge, Massachusetts, and representing the United States, although he was not the first to re-enter the stadium. This was Fred Lors of New York City, who was immediately disqualified when it was revealed that he had travelled several miles by car. "I just got tired," he reported.

According to the *Chicago Daily Tribune*, Hicks had kept up a steady pace until, within two miles of the "goal line," he slowed to a walk whilst climbing a hill, the first time he had ceased running. But the truth seems to be that several miles earlier, he had been on the point of collapse and had begged to lie down. Instead, his trainers gave him a dose of strychnine sulphate and raw egg white, together with several shots of brandy. As a result, he had to be carried over the finishing line and nearly died. A photograph of him taken just after the race shows him with a ghastly and other-worldly facial expression that was probably caused mostly by exhaustion, but which also seems to owe something to the *Risus sardonicus* of severe strychnine intoxication.

Another contestant who possibly had some similar help was William R. Garcia of San Francisco, who was discovered some hours after the race had finished, lying unconscious in the darkness by the side of the course. In this era, the drugging of contestants was not illegal though many considered it unsporting. It seems that all, or nearly all, of the runners were under the influence of one or more drugs, and that strychnine was the drug of the moment.

Strychnine's medicinal career followed a parabolic path. After its initial introduction to therapy there was a rapid increase in popularity and widespread use for a hundred years or so. Then there was a decline as the dubious and modest benefits it conferred were continually reassessed against its severe dangers. The medicinal dose of strychnine was 1 to 3 milligrams but as little as 5 milligrams accidentally rubbed into the eye has produced distinct symptoms of poisoning.

Nevertheless, this powerful nerve poison, like many other nostrums, migrated from the consulting room to the bottles on sale in the pharmacy. Like other drugs once thought to provide a source of vitality to the body, it was later found that strychnine, as a blocker of nerve transmission, is, like alcohol, a depressant, not a stimulant.

Fellows' Compound Syrup of Hypophosphites, sold from an address in Holborn Viaduct, London, cost a stupendous seven shillings for a fifteen-ounce bottle (stamp included; bottle sealed with crimson gelatine and in a watermarked wrapper to prevent substitution). It contained 1/64 grain of strychnine per drachm, later reduced to 1/160 grain. Much of the enormous profit was clearly ploughed into the marketing effort, for there were agents around the globe who could show their contacts a substantial 400-page leather-bound volume of testimonials.[2] Case 1, reasonably enough, was that of Fellows, the inventor: "In the summer of 1864, I was suddenly affected by a copious expectoration of muco-purulent matter...." One started a course of Fellows' syrup with the "tonic dose" of one teaspoonful at each meal, then for more serious afflictions gradually increased this until "the patient experiences the sensation peculiar to the action of strychnine." Some physicians, however, evidently felt that this note of caution was for the faint-hearted. G. Mundie, M.D., L.R.C.S., L.M., L.S.A., writing from Hessel Ease, Yorkshire began his testimonial with the foreboding phrase, "I use your Syrup I fear somewhat extravagantly...."

Easton's syrup was more than twice as strong at 1/30 grains per drachm in 1911, later reduced by half. There are 96 drachms in a pint, so in 1911 a quarter of a pint of Easton's syrup would have been a fatal dose. By the same date, Allen and Hanbury were selling, at 2/6 d (50¢) a bottle, "Byno" hypophosphites, described as "a neutral solution of the hypophosphites of iron, manganese, calcium and potassium with the alkaloids of cinchona and nux vomica, in combination with 'Bynin' liquid malt—a popular tonic which stimulates the appetite, aids the digestion, strengthens and invigorates the nervous system, restores tone to the muscles of the arteries and heart, and acts both as a blood-producing agent and as a concentrated nutrient." A later product, available from about 1930, was Metatone, sold by the American company Parke Davis. It originally contained 1/25 grain of strychnine per fluid ounce together with hypophosphites and other ingredients. Metatone is still sold by their successor, Warner Lambert healthcare, but since the 1970s no longer contains strychnine.[3]

In the United States, vast quantities of patent "bitters," claimed to improve the digestion, purify the blood and so on, were sold throughout the nineteenth century. Although the empty bottles are enthusiastically collected, with about 3,000 different brands being documented,[4] there is little information about what they contained when full, except that it might have been almost anything. Most brands would have contained several ingredients, and it is significant that the label for C.K. Wilson's Original Compound Wa-Hoo Bitters, produced by the Old Indian Medicine Co. of Toledo, Ohio, says that they are "free from all pernicious drugs such as morphine, heroin, cocaine, nux vomica or strychnine." After the turn of the century, there was a considerable tightening of drug legislation, and U.S. Court Judgment No. 4523 of February 1917 found this product to be "a watery solution with sweetening, epsom salts, sassafras and prickly ash," fining the company $25 plus costs. The bitters trade was only a subset of the patent medicine industry. Empty pill tins are also enthusiastically collected, and illustrations of some of these given in a collectors' handbook clearly state strychnine as an ingredient. Wells' tablets, sold by the National Drug Company of Washington D.C. at 50 cents for 100 tablets were made of nux vomica, podophyllin, calcium and ginger and contained 1/500 grain strychnine per tablet ("take two tablets on retiring").[5] "Especially is it to be deprecated that the laity have learned to use these ready-made tablets and compounded pills of strychnine as a constant self-treatment as a laxative," a physician says in 1904. "When persons without discriminative care or professional knowledge turn to the bottle of strychnine, belladonna and aloin or similar pellets...it is as a child playing with fire over a barrel of gunpowder."[6]

Strychnine's parabola can be traced through successive editions of the British Pharmacopoeia. The 1934 edition waxes lyrical about its supposed benefits. It increases the appetite and augments the flow of gastric juice, it says. Reflexes are exaggerated and muscle tone increased. The medulla is stimulated, the movements of respiration are deepened and quickened, peripheral muscles are constricted through stimulation of the vasomotor centre, and the heart slowed through excitation of the vagal centre. The senses are rendered more acute. The blood pressure is raised, so that strychnine improves the pulse and is a valuable tonic to the circulatory system in cardiac failure. It is of value in such conditions as chronic constipation and atony of the bladder, and is much used as a gastric tonic in dyspepsia and as a general tonic in convalescence from acute disease, in treatment of surgical shock, cardiac failure and so on. Given hypodermically in the form of strychnine nitrate in doses varying from 4 milligrams (1/16 grain) up to as much as 67 milligrams (1½ grains) it has been used with success in the treatment of snake bite.[7]

This 1934 description represents the apotheosis of strychnine's supposed beneficial effects, but most of the information had been carried forward from earlier editions.

Twenty-five years later the entry had shrunk to about one-third of its previous length. Small doses of strychnine delay the onset of fatigue, the 1959 edition says, but this is followed by a depression of muscular activity. Owing to its bitter taste it is used as a stomachic and tonic, but there is no evidence to show that it is of any value for these purposes. Neither is it of much use as a laxative nor as an analeptic agent, and it has now been largely abandoned in therapy.

In December 1964, the Council of the Pharmaceutical Society deleted strychnine from the pharmacopoeia for all medicines such as purgatives, although it remained in some tonics for a few more years. It is still used as a neurological research tool. Using strychnine to block the nerve receptors can make it possible to study the physiology of pain causation in experimental animals without actually causing them pain. For a time, strychnine was thought to be effective against nonketonic hypoglycaemia in newborn infants, but this, like other newly claimed therapeutic uses, did not survive the test of continued investigation. By 1971, it was being argued in the *British Medical Journal* that there was no justification for its inclusion in any human medicine, and at about this time it disappeared from mainstream therapy. But strychnine is still one of the drugs routinely tested for in athletes' urine, for the fact that the medical profession has decided that it is too dangerous and ineffectual for medical use does not prevent its occasional reappearance as an illegal stimulant. As recently as 2001, diminutive Indian weightlifter Kunjarani Devi was banned for six months and made to return the gold medal she had won in the 48-kilogram class of the Asian weightlifting champion-ships because strychnine had been detected in a urine sample. "Someone is playing games," was her explanation for the forensic result, according to the *Hindustan Times* reporter. Homeopathic remedies containing small amounts of nux vomica are widely available in India, and she could have taken an overenthusiastic dose of one of these. But of thirty male weightlifters who turned up at the same trials, twenty-two panicked and went home when they were told that they would be tested for drugs.

Nux vomica was, and remains, a mainstay of homeopathic medicine. Such a powerful poison, which seems to destroy the whole functioning of the organism, is ideally suited to the role of universal panacea when subjected to the homeopathic method of giving drugs in such vanishingly small doses that nothing remains. In the words of one among many homepathic websites, "This is the king of our polychrest remedies...a unique gift from the vegetable kingdom...it extends from pole to pole and encom-passes a variety of character that is truly astonishing...especially suited to persons who are thin, nervous, extremely susceptible to external impressions, dark-haired, debauched, given to sedentary life, jealous, hypochondriac males. [Cures] anger, coffee, alcohol, masturbation, sexual excess, injury, hot medicines, rich food, night keeping (?), mental exertion, sitting on cold stones..." and so on for several pages.[8]

Most naturopaths maintain that natural plant extracts exert effects which are different from those shown by the individual purified drugs extracted from them. This may be true and is an area of current drug research. It is certainly true in the case of nux vomica when compared with pure strychnine; for reasons that have already been given, nux vomica is much more danger-ous. To say, as a 1904 author[9] did, that "The natural Nux com-pound, whatever it may be, exhibits a kindliness (!)...impossible to parallel by means of either of the quick-shocking artificial substances that result from the chemical destruction of [its] natural structure" is tripe.

Even greater tripe was propounded by Johann Rademacher (1772–1850), a disciple of Hahnemann, the inventor of homeopathy. According to him, there was no distinction between the disease and the drug. Rademacher was obviously a great strychnine enthusiast, for he called liver disease "nux vomica strychnia." As the poet Philip John Bailey wrote, "The first and worst of all frauds is to cheat oneself."

Is strychnine an aphrodisiac? Traditionally it had such a reputation in India. Its use certainly heightens the senses, particularly the visual, and this seems to be the basis on which nux vomica gained a reputation in Indian medicine as an enhancer of the sexual experience, "not only among Hindus, but also among Mahomedans." The increased sensory perception, producing something akin to the effects of alcohol, caused distillers there to add it to arrack to improve its intoxicating power. It was said in 1870, "The more debauched among the Rajpoots of the province of Guzerat use the nux vomica as a stimulus, but the practice is not general."

In 1853 the magistrate of Goruckpore forwarded the contents of the stomach of one Mungoo, together with a letter from Dr. Aitchison, the civil surgeon, to the following effect: on June 1st, at 9 a.m., the man drank off a bottle of common bazaar spirit, which he had purchased the night before at the Goruckpore Bazaar. Immediately after drinking the liquid, he remarked to a bystander, "Something is wrong; this shurab is intensely bitter." Soon afterwards, spasms came on.... The liquor, which caused death, cost only three pice, whence Dr. Aitchison inferred that it could not have been very strong or intoxicating.... More than one suspicious cause of death, occurring shortly after drinking bitter spirit, will be found in the Nizamut Reports.[10]

The only clear reference to any direct influence on sexual performance comes from some very early studies in France, and nineteenth-century coyness means that the outline of its supposed powers is sketchy. Trousseau and Pidoux claimed that the muscles of the penis were influenced, so that "nocturnal and diurnal erections become inconvenient even in those who for some time before had lost somewhat of their virility." A roofer, age 40, had much weakness of the lower limbs and had not been able to have connection with his wife for seven months. After two weeks of nux vomica extract he walked more firmly, with his genital organs in a state of excitement. A young man, 25 years old, who had been married for 18 months without having any other than "almost fraternal communication" with his wife, acquired his virility under its influence, but lost it again after leaving off its employment. Females too, according to Trousseau and Pidoux, experienced more energetic venereal desires. They have "received confidential information on this point, which cannot be doubted."[11]

Although the enthusiastic 1934 British Pharmacopoeia report makes passing reference to this property, orthodox medical opinion tended to the view that such effects were minor and not to be trifled with in view of the dangers. Let us draw a veil over the nocturnal activities of nineteenth century pre-Viagra society, for they would have wished no less of us, and in any case appear to have left no further information on the subject.

Lack of evidence did not prevent the marketing in 1966 of strychnine as an aphrodisiac by All Products Unlimited Inc., of Miami, Florida, in a product called "Jems," described as "sex energiser nature pep tablets for married men and women." The company was indicted for conducting a scheme to obtain money through the mails in contravention of U.S. code 39/4005. This prosecution was not because strychnine was dangerous, but on the grounds, which the president of the company admitted, that it had no proven sexual function and the appeal of the product was entirely psychological. Any effect on the lower spinal nerves, according to the expert witnesses consulted, was toxic not stimulatory. It seems a pity that the company caved in on the actual effects of strychnine and did not call any medical witnesses to cite Trousseau and Pidoux. On the admitted false representations, a fraudulent order was issued by the court. This being the 1960s, still less was there any criticism of the company's marketing strategy, which was to place advertisements in the "colored" newspapers... "to add strength and night pep in your home life," having discovered that "the word 'Nature' to colored people means their sex organ."[12]

We are faced with a large discrepancy when considering the nineteenth-century trade in nux vomica. The mystery springs from the following passage, dated 1882.[13]

> The seeds, or beans as they are called, chiefly come from the East Indies. Forty years ago the extent of the importation was only about 600 lbs; it now amounts to 6000 cwts. In consequence of these large importations of nux vomica it has been thought that it was used by brewers to give a bitter taste to ales. This, however, has been disproved, but as the consumption for medicinal purposes is but small, it is still unknown to what use the bulk is put.

The *Encyclopaedia Britannica* of 1884 gives a similar figure, 200 tons of imported seeds, valued at £1600. If these figures are to be believed, importation went up by a staggering 112,000% during the four decades from 1840 to 1880. Smith's earlier figure seems to be incorrect, for the official customs returns show that by 1819, importation (including re-exports) was already 17 tons.[14] But the growth over the years was still enormous. Flückiger and Hanbury, writing in 1879, gave even greater figures.[15] "We have seen 1136 packages offered in a single drug sale on March 30, 1871. Exports from Bombay were 3341 hundredweight (cwt), from Madras 4805 cwt, and from Calcutta 2801 cwt, all shipped to the United Kingdom"; a total of over 10,000 cwt, or 500 tons. Another review published in 1904 refers to the worldwide demand for the seeds being even greater: "Immense quantities, thousands of tons." By this time, the market at last became saturated so that the price of nux vomica seeds on the London market collapsed to 7s–10s per cwt, which did not allow their profitable collection.[16]

Assuming an average strychnine content in the nux vomica of 1%, even Flückiger and Hanbury's figures mean an importation of five *tons* of pure strychnine per annum. If the human fatal dose is, for the sake of argument, 50 milligrams, this is enough to fatally poison 100 million people, the entire population of Europe at that time, even neglecting the brucine. Where did it all go?

A letter to *The Lancet* as long ago as 1830 throws only a small glimmer of light onto the subject. J.H.G. of Bath, full name not given, clearly has some disturbing suspicions.

Now, Sir, I wish to ask...what becomes of this enormous quantity of poison; a trifling quantity is employed in medicine, and for the destruction of vermin, and by poachers we know that a comparatively small quantity of these articles is also used; but if there be any other open, honest or avowed purpose to which they are applied in any part of the world, it would remove some very unpleasant suspicions and be very gratifying to many persons to be made acquainted with it.[17]

The letter never elicited a reply, or if it did, it was never published. The answers to J.H.G.'s question involve pushing a broom into some corners of Victorian life that are very murky indeed.

Were the gamekeepers of the kingdom responsible for the growth in nux vomica use? J.H.G. appears to think not. Strychnine certainly became one among many products put at their disposal during the nineteenth century. Obtaining any kind of quantitative information is difficult, but as far as can be seen, the poisoning of vermin would never have accounted for more than a small proportion of the imports. Chevallier[18] says that about 600 kilograms a year was used in India during the 1870s to poison large animals, "tigers, lions, etc." It was also used in France against foxes, but quantities are not given.

The Gamekeeper, in its first full year of publication in 1892, contains no advertisements for strychnine-containing pesticides, and only one or two slight mentions advising correspondents how to use it to deal with particularly serious rat infestations. The unwillingness to use strychnine is certainly not out of squeamishness, for there is much discussion of other methods for killing rats using drowning-traps and so forth. Strychnine was just too dangerous for widespread use where valuable sources of food such as rabbits might become contaminated. In 1856 the procurator fiscal in Scotland warned gamekeepers that they faced criminal prosecution for any deaths or injuries caused as a result of using strychnine. At the trial of Henry Bowles in 1888, it was said that strychnine was "occasionally used by gamekeepers in Staffordshire and Derbyshire for killing vermin," while in *Times*'s account of the trial of Walter Horsford in 1898, a police inspector who visited various chemists' shops is reported as saying, "The records of chemists in St. Ives, Huntingdon and St. Neots contained no record of any sales of strychnine."[19] If Ethel Major's father in

Chapter 17 of this book is in any way typical, gamekeepers had (or should have had; Bowles was the exception) a healthy respect for the poison, and kept their small amounts of it under lock and key.[20] Use of strychnine on a large scale was rare enough to be noteworthy. Much later, in 1927, there was a Hamelin-like plague of mice in California, caused by bumper crops and the elimination of many of their natural predators such as badgers and snakes. The appropriately named S.E. Piper of the U.S. Biological Survey solved the problem in a month with the aid of 40 tons of strychnine-laced alfalfa. But this was a newsworthy event.[21]

In Victorian times, sweets were often coloured with arsenical pigments. Cake decorations were particularly poisonous; two or three iced cakes were frequently a fatal dose, and there were numerous fatalities.[22] Not surprisingly, there were frequent scares about the adulteration of food and drink. At first, statutory controls were virtually nonexistent, and neither were there yet the analytical methods needed for the reliable detection of adulterants. The foundation of the Society of Public Analysts in 1874 was symptomatic of the desperate need to effectively enforce legislation which by then had already been passed. Of twenty-six samples of curry powder purchased in London in the early 1850s, nineteen were found to contain red lead.

The peak of one scare was reached in 1851–1852. Since the 1820s, the Allsopps of Burton-on-Trent had been manufacturing India Pale Ale, an especially bitter ale originally brewed to survive long enough in bottle to be shipped out to India in ships returning from the spice run.[23] In 1851, rumours began circulating that Allsopp's ales were being adulterated with strychnine to increase their bitterness. As a result, Mr. Henry Allsopp felt it necessary to commission at considerable expense an independent report by Thomas Graham and A.W. Hofmann, the two most prominent chemists in the country.[24]

Graham and Hofmann firstly showed by experiment that the amount of strychnine that would have to be added to beer to replicate the observed bitterness of India Pale Ale was one grain per gallon, or twice the fatal dose. The amount required to adulterate the 200,000 barrels of Burton India Pale Ale brewed each year would amount to 16,448 ounces at a cost of £13,158; many times the current world production of strychnine, which they put at less than 1,000 ounces. Secondly, the quality of bitterness imparted to beer by strychnine was distinctively harsh, metallic and persistent, quite unlike the pleasant aromatic bitterness of hops. And thirdly, they developed a simple analytical method for isolating strychnine from beer and tested numerous samples of Allsopp's ale from various London bottlers. All tested negative. (The samples were bottled before the rumours circulated.) They concluded by saying, "No one who has witnessed the open manner and gigantic scale in which the operations are conducted in [Allsopp's] establishment, as we have done, could entertain the idea for a moment that any practice involving concealment was possible." They pointed out that had the brewers wanted a cheap alternative to hops, the harmless and agreeable quassia bitters would have provided a much better alternative than strychnine. This remark in itself indicates how much leeway Victorian brewers had to put almost anything they liked in their beer without announcing it.[25]

Justus von Liebig, the most famous chemist in the world at the time, who had advised the Burton breweries in years past, added his support. Hoffmann sent a crate of Allsopp's ale to Germany via the Continental Delivery Company. Liebig carried out a subsidiary investigation which consisted of drinking a bottle, and declaring it excellent. For a payment of £100 from the brewers, he wrote a public testimonial, which arrived in a pleasing early form of Denglish:

> The wholesomeness of the Pale Ale as a general beverage, both for invalids as well as for the robust hat diesem bier mit Recht die Gunst der Ärzte und das Publicums zugewendet...durch eine besondere Untersuchung von Pale Ale aus Allsopps Brauerei aus verschiedenen stores of the London bottlers genommen...

Copies of the report, suitably Anglicised, were put up in omnibuses and railway carriages and the *Medical Times* withdrew its allegations.

So far, so good, but the investigation leaves more questions unresolved than it answers. The report was financed by Allsopp, and dealt only with Allsopp's ales. Bass, another famous brewer, independently commissioned a report from Thomas Wakley.[26]

Other smaller brewers might not have been so scrupulous. But more importantly, we have seen how the similarly bitter brucine was a much cheaper commodity than strychnine—literally a "drug on the market"—and they do not even mention it.

How did the rumours start? According to Graham and Hoffmann, it all came from Pelletier, the scientist who first isolated pure strychnine. A scientific colleague of Pelletier's, Payen, had said in a public lecture in Paris, that Pelletier's manufacturary had once received an order for an extraordinary amount of strychnine, for a destination which on enquiry, Pelletier had found to be a brewery in England. The *Medical Times* journalist had then made a story out of it. This remark of Payen's, which referred to events of ten or twelve years before, could not be further substantiated because Pelletier had died in 1842. Payen did say, however, firstly that an order of fifty or a hundred ounces would have been considered "large" by Pelletier, and secondly that the manufacture of strychnine in France had not increased in the subsequent years up to 1852. Payen defended his public statement by pointing to a note in Chevallier's book[27] saying the same thing. In fact, the information given by Chevallier is equally tantalizing. The 1850 edition in one place lists a dozen or so known adulterants of beer and states unequivocally that certain brewers in England and France have added nux vomica powder or Saint Ignatius beans, but in another place he mentions that it is "said" to have been added, but that this has not been unequivocally demonstrated. This information was carried over unchanged to Chevallier's 1875 edition. An 1872 textbook of organic chemistry repeats the charge and says unambiguously that there have been fatalities.[28] In Germany, there had been a prosecution of a Westphalian brewer as long ago as 1825.[29]

Graham and Hofmann's figure for world annual strychnine production of 1,000 ounces is far too small. It may have represented the true figure for pharmaceutical grade strychnine handled by reputable druggists. Other sources make it clear that beer adulteration was almost universal. A cartoon by (probably Isaac) Cruikshank as early as 1807 shows a brewer shepherding several diminutive demons, one of them labeled "nux vomica" into his barrel: "Come along my little boys, that's my darlings." John

Bull, looking on, comments, "Od lookers, what be I to take all those fellows in my guts, why I shall ne'er want any more Physic." Nearly forty years later, the cartoon reproduced in Figure 17 shows that nothing had changed.

It was the publicans, not the large brewers, who adulterated beer. "We have met with instances in which an article which had been denominated beer, has been produced without containing a particle of either malt or hops."[30] This adulteration became widespread during the Napoleonic wars, and was instigated by a certain Jackson, a chemist. The publicans sold their beer to the customer at the same price at which they bought it from the brewer, so all their profit came from adding water together with various other ingredients to restore the bitterness, intoxicating properties and foaming head. Before long, carts could be seen in the streets brazenly bearing the brightly painted inscription "Brewers' Druggist," despite legislation from the previous century forbidding the addition of any adulterants. Hop production hardly increased in the thirty years to 1855, but in the decade from 1843 to 1853 alone, beer consumption went up 34%. By the middle of the nineteenth century, nux vomica, under the name Fabia Amara,[31] was a frequent adulterant, alongside other "bitters" such as gentian, quassia and the equally highly toxic *Cocculus indicus*. A recipe given by a Mr. Morris in his book *Brewing Malt Liquors* was: malt, 4 quarters; *Cocculus indicus* berry, 4 lbs, sugar, 28 lbs, Fabia Amara, 6 lbs. According to evidence given before a parliamentary committee, the publicans would adulterate their beer more on Saturday nights, so that Sunday became notorious as a day of diarrhoea. In Germany in 1862, a man called Creuzberg evaporated publicly available beer onto corn meal and fed it to hungry dogs. All were poisoned, and several of them died.[32]

Government officials seemed oblivious of the scale of this trade and appear to have been ignorant of the difference between strychnine, nux vomica and brucine. George Phillips, chief officer of the chemical department at the Board of Inland Revenue, told a Commons Committee that there was no foundation to the rumours about strychnine in beer. "I do not think there is a large quantity of strychnine brought into the country. I know that when I want any I have frequently to send to Paris for it."[33] Other, more percipient witnesses pointed out the enormous deficit to the revenue from the loss of tax on adulterated beer, for the 200 tons of nux vomica that was annually imported from India attracted a duty of only two shillings a hundredweight.

Also, as Krätz points out, why Allsopp and Bass, and no other brewers? Why did Allsopp feel that he had to take up the cudgels so energetically if the rumours concerned the beer trade generally? Could the truth be that adulteration was universal but the two largest and richest brewers were the easiest targets for a very early manifestation of corporate blackmail? If so, no evidence has survived to support this hypothesis, which is hardly surprising. As to who might have been blackmailing them, surely it is fanciful to table the observation that Burton is only thirteen miles from Rugeley.

Abortion appears always to have been widespread in India: "This detestable crime prevails to an almost incredible extent" ...coming in the wake of "The extreme laxity of morals and indelicate conversation that is universally indulged in amongst Hindoo families," although nux vomica is not included in the author's (extremely short) list of plant drugs used to procure abortions on the subcontinent.[34] But it is reported to have been used extensively by the Chinese for procuring abortions, by the introduction of nux vomica powder into the vagina.[35]

From the 1840s on, there was a large increase in abortion in the Western world, particularly the United States; the abortifacient drug industry emerged as a large-scale business. A visiting French physician writing in 1867 was shocked at the "licentiousness" of advertisements such as those for "French Lunar Pills," prominently advertised in the newspapers for "eradicating all impurities thereby assisting nature in performing its office for ladies in a delicate health." Some of the advertisements made the purpose clear, while at the same time attempting to protect the advertiser from prosecution, by carrying a warning that the preparations could be safely used at all times except during early pregnancy, when a miscarriage might result. A Detroit doctor noted in 1874 that a single supplier in the city had sold over 500 lb of pills in one year.

Medicine at this time had no preparations capable of reliably inducing abortions, although drugs such as strong laxatives might produce terminations in women prone to miscarry. However, physicians and their patients thought that there were such agents. Information on the contents of the pills is understandably very hard to come by; none of the firms would have kept documentary records, and analytical techniques were primitive. In 1869 the intrepid Ely van der Warker bought samples of eleven different pills available in New York and tried to determine their content by taking them himself. Some seemed inert, some contained laxatives, but others appeared to contain dangerous drugs such as black hellebore and ergot. Other herbs known to have been used included pennyroyal (the reputation of which had spread from England), black cohosh, cottonroot and many others. Nux vomica, because of its very powerful physiological effects, would have been among these. Van der Warker wrote at the time, "Druggists do not trifle with their reputations as skillful abortionists. The tenacity with which even respectable druggists will sell violent and noxious drugs to women far advanced in pregnancy forms one of the most alarming features of this trade."[36]

The Encylopaedia of Advertising Tins illustrates a number of small enamel tins for drugs sold openly or given away as samples to physicians. The tin for Prairie Blossom Pills claims that the pills "Make Women Beautiful and Strong" and "Cure While She Sleeps." There were also Dr. Martel's French Female Pills and Du Bois Pacific Pills. These can easily be distinguished from the other products because of their high price: $1 or $2 a tin at a time when laxative pills and others typically cost 10 cents.

At the time, the association of a death-dealing material with the topic, sex, that raised so many unspoken anxieties, was impossible for open discussion. Here is the court record of the way in which the topic of abortion was raised during the Cream trial, during the course of the examination of Dr. Stevenson, the forensic witness. M'Culloch, it will be remembered, was the Canadian whom Cream befriended during his final visit to North America.

> Mr. Geoghegan handed to his lordship a question in writing, in order to ascertain whether it was desirable that it should be put to Dr. Stevenson. Mr. Justice Hawkins said that, so far as the nature of the question went, they could not consider that, but in a court of justice they were bound to ascertain anything that might bear on the case.

Mr. Geoghegan: Quite so my lord. Those persons who come into a court of justice must be prepared to hear things that may shock them. It is only for consideration of public morality that I have submitted the matter to your lordship. (The question was then handed to the witness.)

Dr. Stevenson: I have read it, my lord, yes.

Mr. Geoghegan: But it is essential that the jury should know the question my lord, as well as the answer.

Mr. Justice Hawkins (handing the question to the jury): Read it, Gentlemen. You will see the obvious grounds on which I think it expedient that it should not be put in words.

Mr. Geoghegan: It is with reference to the evidence of M'Culloch that the question is put. Is that an American theory?

Dr. Stevenson: It is a matter of common medical knowledge in this country. It is supposed to have come from America. It was mentioned in a notorious pamphlet.[37]

There were a few legitimate commercial uses for the nux vomica alkaloids. Brucine, being completely odourless but extremely bitter, has been (and may still be) added in traces to perfumes, all of which contain alcohol, to denature them,[38] so that anyone who has ever nuzzled the neck of an elegantly prepared woman has known the taste of its sinister allure. This use would probably not have come in until 1900 or later. Liebig[39] thought that the use of strychnine for domestic poisoning of rats and mice was enough to account for the increased imports during the previous century. But the strong suspicion remains that there was a large backstreet commerce in commercial-grade nux vomica, strychnine and brucine. It is remarkable how no one, including the government, seemed to have firm information, or even much interest in obtaining firm information, in the trafficking of such dangerous substances. Further research may one day turn up more details about this nefarious trade.

Endnotes

1. *Chicago Daily Tribune*, 31 August, 1904; *New York Times*, same date.
2. Fellows, 1881.

3. An acquaintance of mine went to the doctor in about 1965 and complained of light-headedness and other strange symptoms. The doctor questioned him and found him to be drinking a bottle of Metatone a day. His name, coincidentally, was Palmer.
4. Ring, 1980–84.
5. Zimmerman, 1994: 166.
6. Howes, 1904.
7. This contains an arithmetical error. 67 milligrams is approximately one grain not 1½ grains. Either dose is potentially fatal.
8. Patel, J.D., Gems of Homeopathic Materia Medica, www.indiangyan. com/books/homeopathybooks.shtml.
9. Howes, 1904.
10. Chevers, 1870.
11. Pereira, 1855; Trousseau and Pidoux, 1839: 233.
12. www.usps.com/judicial/1966deci/2-157.htm.
13. Smith, 1882.
14. Kew Public Record Office, document CUST 5/8. The 1856 return shows a small amount imported from New South Wales.
15. Flückiger and Hanbury, 1879
16. Talbot, 1911
17. *Lancet*, 1830:678
18. Chevallier, 1875
19. Although he goes on to say, apparently contradicting himself, that " …in the poisons registers which he searched he found considerable quantities of pure strychnine as being purchased by farmers."
20. It is notable that the majority of strychnine poisoners, beginning with Palmer, came from the North of England, in a belt stretching from Huntingdonshire to County Durham. Was this to do with easier availability of the poison, or to a kind of local tradition? (In the early years of the century, there was an outbreak of arsenical poisoning in East Anglia.) One exception was Bowles (Surrey), but he and his wife had moved there from Derbyshire, and had probably brought the strychnine with them. There were one or two cases in London and the South, but the North was heavily over-represented.
21. Duke, 1985.
22. Chevallier, 1875.
23. Tomlinson, 1994.
24. Graham and Hofmann, 1852; The Times, 7 June 1852:7.
25. Lord Campbell, when practising at the bar in 1820, defended someone in an action for defamation who had said that a brewer boiled toads in his beer. Campbell laconically quoted Macbeth to prove that even if he did, it would just add character. The brewer was awarded one shilling damages (Campbell, 1881:I:381).
26. Hassall, 1876.
27. Chevallier, 1850.
28. Berthelot, 1872.
29. Mitchell and others, 1855.
30. Accum, 1820.

31. Adulterants often had cryptic pseudonyms. Sawdust was called "Powder of post."
32. Krätz, 1990, Otto, *Chemie in Unserer Zeit*, 24:3.
33. Mitchell and others, 1855.
34. Shortt, John, 1867. *Trans. Obstetrical Soc. London*, 9:6
35. Duke, 1985.
36. Mohr, 1978.
37. Shore, 1923.
38. That is, to stop alcoholics drinking them, and to save customs duty on the alcohol content. Brucine may still be used for this purpose in some parts of the world, although the more usual denaturant these days is the similarly bitter synthetic compound denatonium benzoate or bitrex. (I am grateful to Dr. Roger Snowden of Firmenich SA for this information.)
39. *Times*, 7 June 1852:7.

CHAPTER 15

The Blue Anchor Murder and Other Outrages

My pulse will be quickenin'
With each drop of strych-a-nine
We feed to a pigeon
It just takes a smidgen
To poison a pigeon in the park!

Tom Lehrer, *An Evening Wasted*
with Tom Lehrer, 1959

Scientifically speaking, we left strychnine and brucine in the hands of Pelletier and Caventou back in the dawning years of the nineteenth century. Subsequently, the two alkaloids played an important part in the history of science, and now is the time to bring the story up to date. In terms of their chemistry, the two alkaloids remained intractable enigmas for well over a century. They were even given their own quasi-alchemical symbols,[1] St⁺ and B⁺. It was, or should have been, a cause of constant uneasiness to the nineteenth-century chemists that the fate of a suspected murderer could depend on the correct interpretation of the results of colour tests on substances about which virtually nothing was truly known.

The first step in establishing the nature of an organic drug is to determine by analysis the number and proportions of the atoms making up the molecule. This was a difficult task. Numerous

221

analyses of strychnine, brucine and their salts were carried out. The first, by Pelletier and Caventou, were not very accurate, but by the 1830s the correct molecular formula of strychnine, $C_{21}H_{22}N_2O_2$, had been determined. It then took more than a century of continuous work to find out the chemical structure, that is the way in which those forty-seven atoms are joined together. It was an enterprise of virtually unparalleled persistence, requiring people of exceptional ability. The techniques that evolved during the century-long quest for the solution of the strychnine structure would have widespread applications in many other fields of chemistry.

The first serious attempts to climb the foothills of strychnine chemistry were made from 1886 onwards by three groups of German chemists, although a large number of desultory experiments had been made in the previous decades. The strychnine molecule was highly stable and resistant to degradation. This meant that if it were treated with harsh reagents under extremely forcing conditions, some identifiable products could be obtained, but these were small fragments resulting from the almost complete destruction of the molecule. For example, heating strychnine strongly with zinc dust gave ethylene (C_2H_4), acetylene (C_2H_2), hydrogen, ammonia and the only slightly larger molecule, carbazole ($C_{12}H_9N$). Did this mean that strychnine contained a carbazole nucleus, or were the conditions so harsh that a rearrangement had taken place during the degradation? The new, more systematic, attempts lasting into the early years of the twentieth century only served to emphasise the difficulty of the problem.

Robert Robinson was born near Chesterfield, an industrial town on the edge of the peak district of northern England, on September 13, 1886, into a middle-class congregationalist family of manufacturers.[2] His great-grandfather had established a business making surgical supplies. His father married twice and sired a total of thirteen children, of whom Robert was the eldest of the five by his second wife.

As a child, Robert was keen on mathematics; he later became a national-level chess player. But his father wished him to contribute to the family business, and so he went to Manchester University to read chemistry. Manchester in 1902 was the leading centre of chemical research in Britain, and with W.H. Perkin, Jr. (the son of the discoverer of mauve and founder of the British dye indus try) at its head, it alone could rival the attainments of the great German universities. Initially, Robinson was a moderate student, but teaching by Perkin proved inspirational. After graduating, he stayed on as Perkin's research student, during a period of achievement that lasted until Perkin moved to Oxford not long before the First World War.

Another luminary of the Manchester department at this time was Chaim Weizmann, who later became the first president of Israel; while at Manchester, Weizmann rescued the Allied war cause and effectively founded the biotechnology industry by showing how acetone for munitions could be manufactured using bacteria. Another colleague was Holger Erdman from Stockholm, the founder of chemotaxonomy, the science of working out the relationships between plants by studying the natural products they produced. The botanists, used to using anatomical characteristics such as flower shape, were highly sceptical; but as we have seen with the genus *Strychnos*, the germ of the idea goes back at least to Linnaeus.

Robinson was an intellectual and physical workhorse (he used to work in the laboratory until 3:30 in the morning) with mental tenacity and an encyclopedic knowledge base that made his mind a fertile ground for inspiration. He was eccentric and difficult. Even his "official" obituary[3] feels obliged to refer to "remoteness verging on rudeness" and of his being "wholly emotional in his immediate reaction to people and things" in a way that could be frightening. He hardly ever looked at the person he was talking to, a habit that caused immense problems to one of his later research students, who was deaf. R.N. Chakravarti, who came to Oxford from India after the Second World War to work on strychnine, described him as "a very moody and peculiar type of person."[4] He could be a terrible lecturer, getting lost in his own thoughts so that only the very ablest students had any idea what he was talking about. But others recorded his "very remarkable personal charm," and how he could be "jovial and great fun."

There was a quirky restlessness in him that may have arisen from feelings that the world did not take him sufficiently seriously, despite his self-evident preeminence in his own field. Early on, he moved departments frequently and some of these changes probably resulted from interpersonal friction. There was certainly plenty of this during his only career venture outside the university laboratory. Following the exposure of the inadequacies of the British chemical industry during the First World War, he became director of research at the British Dyestuffs Corporation, the product of a government-funded amalgamation of private companies. This unsuccessful interlude was marked by intense infighting, and his rapid return to university life. Neither was there great harmony following his eventual arrival at Oxford in 1930 at the age of 44, despite the fact that he was succeeding his mentor Perkin, who had died the previous year. He stayed there until retiring, but did not appreciate the collegiate system. The custom for the most junior fellows to pass around the wine and dessert at Senior Common Room dinners Robinson—who had already held professorships at five other universities (Sydney, Liverpool, St. Andrews, Manchester and London)—considered beneath his dignity. Although he was an enthusiastic amateur mountaineer, at one point he smoked one hundred cigarettes a day, and drove the short distance from his house in North Oxford to the Clarendon Laboratory in a huge Studebaker, which he and his wife Gertrude drove and maintained very badly, invariably forgetting to inflate the tyres. After Lady Robinson—organic chemist, academic hostess and pillar of Oxford respectability—died, he remarried at the age of 71 a Hungarian-American blonde 25 years his junior, whom he met on the Queen Mary. She turned out, perhaps not surprisingly, to be something of a gold digger, causing him to fear "a wreck on the rocks of luxury girlmanship," caused partly by her "readiness to bet on two flies crawling up a window-pane."[2]

There is no denying the brilliance of Robinson's work. In 1917 his fame was established when he disclosed a remarkable synthesis of the alkaloid tropinone (a relative of cocaine). The structure of tropinone had been proved in 1901 by the German chemist Richard Willstätter of the University of Munich, who had synthesised it in twenty immensely laborious stages. Two years after Willstätter won the Nobel prize in 1915 for his outstanding synthetic achievements, the young Robinson made tropinone by mixing together three common laboratory chemicals and letting the mixture stand at room temperature in solution for half an hour. The mild conditions, mimicking those present in the plant, and the "inevitability" of the way in which the components came together, added up to what is admiringly called an elegant synthesis, a triumph of inventive reasoning. Later he corrected the structure of morphine proposed by others, showing how his alternative structure would agree with the way in which the plant would make it, then went on to prove it.

When Perkin and Robinson published their first paper on strychnine in 1910, they made the following comment: "Although during the last fifteen years much systematic and valuable work has been done on both strychnine and brucine by Julius Tafel, Hermann Leuchs and others, it cannot be said that anything of real importance is known as to the structure of these alkaloids." This is not strictly true; for example, in 1884 another chemist Hanssen had converted both strychnine and brucine into the same compound, and thus confirmed that they were closely related. Soon afterwards Perkin and Robinson were the first to publish a suggested structure for the molecule, but they acknowledged that this attempt was no more than a working hypothesis intended to focus further work.

The strychnine problem could have been tailor-made for Robinson, with its combination of apparently almost insuperable difficulty and scope for attack in a multitude of ways. Milder manipulations were able to transform strychnine into a variety of products, all seeming to contain the unchanged basic structure. The Robinson group and others thus built up a sizeable "island" of chemically modified strychnine derivatives of unknown structure that could be interconverted with each other, but not with anything else that was known. As time went by, the island became a veritable continent. Hundreds of reactions were carried out at Oxford. It was not unusual for researchers there to handle 200 grams of pure strychnine at a time, and the chemistry stores in the 1930s must have contained enough of it to poison everyone in the town.[5]

The number of weapons in the arsenal of the investigators in the classical period was strictly limited. They had to try to degrade the molecule to produce smaller fragments using the mildest possible conditions. One fragment might be identical with a substance already known, either synthetically or by degradation of another natural product. If there was no luck at this stage, the degradation product could be subjected in turn to a variety of other reactions, always with the aim of producing something smaller that could be definitely identified. At each stage it was necessary to argue by analogy from the known reactions of simpler substances to try to explain what was happening, and to suggest what approaches might be fruitful for the next stage of degradation. The criteria for validating this work were extremely limited. If two samples obtained by different pathways had the same melting point and other physical properties, this proved that they were the *same*. It did not prove what they *were*. Everything else had to be proved by logic from first principles. The reactions that turned up during these degradations could not always be understood. It was often necessary to synthesise whole series of simpler model compounds to clarify what was happening. To add to the difficulty, some natural product molecules could undergo rearrangements of the carbon skeleton even under mild conditions, although the strychnine skeleton itself is resistant to rearrangement. As Robinson himself wrote later, "The great difficulty of the problem of determining the constitution of strychnine will be realized from a consideration of the great effort which has been required to solve it, especially when it is understood that large quantities of the pure alkaloid have always been readily available for investigation. Furthermore, the bases themselves and many of their derivatives are beautifully crystalline and it cannot be claimed that there has been any unusual difficulty in the manipulations required."[6]

By 1931, when Robinson reviewed progress, the mere outline summary of what was known about strychnine took twenty-one pages of closely written argument, culminating in a new tentative structure that was not only incorrect, but left three of the carbon atoms still unaccounted for. By the following year, the Robinson team was able to advance a full structure proposal. This accounted for all of the carbon atoms and was their proposed solution for several years, including the war period, when work on the problem stopped and the Oxford laboratories were turned over to penicillin and other important strategic work. Leuchs, after publishing 130 papers[7] and without ever seeing a solution to the problem, met his death "under tragic circumstances" during the fall of Berlin in May 1945.

After the war, evidence that even the 1932 structure needed modifying began to accumulate, not all of it from Robinson's laboratory. The correct structure had in fact first been put forward as a possibility by the Robinson group as early as 1934, but the credit for its choice as the final version had to be shared with brilliant new arrivals on the scene, the Yugoslav Vladimir Prelog and the American R.B. Woodward.

Strychnine structures; (a) the "working hypothesis" of Perkin and Robinson, 1910; (b) Robinson's proposal of 1932–1939; (c) Prelog and Szpilfogel, 1945; (d) Robinson's "Frankenstein" structure of 1947; and (e) the final structure (Robinson, Prelog, and Woodward), including stereochemistry as determined by x-ray analysis. Brucine is dimethoxy substituted in the aromatic ring.

So Robinson's triumph was not clean-cut. There was a curious diffidence about the way in which he communicated his results to the world. His main account was published in a new review journal which did not survive more than a few years and is now difficult to obtain. As he says there, "Frequent changes of opinion in the last phases may have given rise to the suspicion that another modification may soon be forthcoming."[8]

But by the 1950s, the chemical structure of the strychnine molecule was at last established. It had taken more than a century and had engaged the skills of some of the finest scientific brains of the era. A back-of-the-envelope calculation suggests that reaching the state of knowledge that *this* arrangement of lines on a piece of paper, representing one particular organisation of atoms in the molecule, rather than *that* alternative arrangement, represented at least 500 man-years of effort. Was it worth it?

Science advances on a broad front with occasional break-throughs. There were no sudden dramatic advances in strychnine chemistry; it was decades of hard slog. The incidental benefits that accrued in terms of new insights, new methods and new ideas by the efforts of Robinson and others working on such a difficult problem over so long were considerable. Whether those talents would have been better used in solving less academic challenges is impossible to answer. Nowadays, his papers from the 1930s seem dusty and remoter in their scientific appeal than Pelletier and Caventou's excitements of more than a century before.

Robinson died in 1975, at the age of 88, handicapped by severe tunnel vision, but still mentally alert. In his last years, he worked on several books simultaneously. His autobiography,[9] of which he completed only the first of two promised volumes, was a disaster. Apart from its entertaining incidental highlights, it is of interest neither to chemists nor to the general reader (who, as Williams points out[10] is asked to follow statements such as "the description 'hetero-enoid' is self-explanatory"). He also planned a book on strychnine, one written to describe the details of his research on the structure problem and to set the record straight about various priority disputes. But it was never written. This book, for better or worse, is not it.

Robinson was not the only man of this era to suffer wife trouble. An American poisoning case of the late 1920s is one that shows how greater scientific knowledge since the Victorian era, coupled with greater professionalism in forensic science, was dramatically improving the chances of unequivocally detecting a strychnine poisoner.

Disabled World War I veteran Carroll Rablen had a little in common with the unfortunate Daniel Stott, who met his death at the hands of his wife Julia and Thomas Cream. Rablen's wife, Eva, seems to have found him an encumbrance. In April 1929 Rablen took her to a dance at the local schoolhouse in Tittletown, California, but being deaf, he unwisely decided not to participate in the bacchanalia but to sit outside in his car while she went inside to enjoy herself; within limits, of course. Around midnight, Eva very kindly brought him out some coffee and sandwiches. His last words to those who rushed out to render assistance as he writhed in agony were that the coffee had tasted bitter, then he died. Eva was distraught. What a piece of luck that she had taken out $3,000 life insurance on him not long before, and even luckier that an analysis by a local chemist failed to find any strychnine in his stomach.

The local sheriff could not feel so genial, though. He ordered a search of the area of the schoolhouse and a broken bottle labelled strychnine and bearing the address of a local pharmacist was found. The pharmacist said that the bottle had been bought by a woman who wanted it to kill gophers, and he identified Eva as the woman. A second analysis was ordered, and forensic expert Dr. Edward Heinrich found strychnine in the stomach and the coffee cup. In a minor stroke of genius, Heinrich reasoned that on her way through the crowded schoolroom out to the car, Eva would almost certainly have spilled some coffee, and appealed to anyone who had been present to examine their clothing. A woman came forward with a stain on her dress, and Heinrich detected strychnine in it. Eva got life imprisonment.[9]

Meanwhile, back across the Atlantic, another case was characterised by the strange and unprecedented claims made on behalf of strychnine by one of the most peculiar individuals ever to take ship across the English Channel.

In January 1924, at the Hotel Victoria in Biarritz, a bizarre relationship sprang up between an Englishwoman and a Frenchman, Jean Pierre Vaquier, who was employed by the hotel as radio operator, charged with providing radio concerts to the assembled guests in the evenings.[12]

The woman's name was Mabel Theresa Jones and she had come to the resort to convalesce after a nervous breakdown caused by the failure of her catering business. What made her liaison with

Vaquier grotesque was that she spoke no French and he no English; but this did not prevent them taking a dictionary to bed with them within a few days. Nor did it prevent him begging her tearfully not to leave him when she received a telegram from her husband Alfred to rejoin him at Byfleet, Surrey, where they had recently purchased the Blue Anchor Hotel. She started back to England, accompanied by Vaquier as far as Paris and spending the night with him there before catching the boat train back to her husband.

Not long afterwards, Vaquier arrived in London, which he had never visited before, and took up residence at the Hotel Russell, ostensibly to try to sell the rights in a sausage machine that he had patented. (A model of it was produced in court.) Here he made the acquaintance of a French-speaking chemist, Mr. Bland, buying from him various harmless chemicals that he said he needed for wireless experiments. Mabel Jones visited him at the hotel and they were seen in bed together by a chambermaid called Annie Muff.

On February 14th, Vaquier suddenly arrived unannounced at the Blue Anchor Hotel and took up residence, staying free, borrowing money from, and apparently living on cordial terms with, both the Joneses. It did not appear that any sexual activity took place between him and Mabel during the six weeks that elapsed before Alfred Jones died.

This occurred on March 29, 1924. The previous night there had been a party at the hotel, with quite a bit of drinking. Jones was a consistently heavy drinker whose post-mortem showed fatty degeneration of the liver, but was otherwise healthy. He kept a bottle of bromo salts on the bar-parlour mantelpiece and invariably helped himself, and others, to a dose of these on the mornings after their drinking bouts. On the Saturday morning in question, Vaquier came down early and would not budge from the bar-parlour although it was very cold and the gas fire was in the next room. No one recalled seeing the salts bottle the night before, but when Jones eventually made his appearance it was there on the mantel shelf, with about two teaspoonsful of contents. He spooned out a dose and swallowed it down in water. "Oh, God, they are bitter," he said, according to Mrs. Jones, and he may have gone to the door and tried to spit some of them out. Mabel Jones found that most of the rest of the contents of the jar were "long, thin crystals," tasted them finding them extremely bitter, and put the bottle away in the back of a drawer. She then mixed her husband an emetic of salt and water, which made him vomit but did not prevent him dying within the hour.

All of this took place in front of several other people, members of staff of the hotel, who seemed to carry on with their usual activities, dusting, scrubbing the doorstep and so on, more or less undisturbed. One of them testified that while Mrs. Jones and the doctor were attending to Mr. Jones upstairs, Vaquier had rushed in, shouting, "Medicine, Doctor, quick," and pointing to a similar bottle of salts belonging to someone else. This prompted her to show Vaquier where Mrs. Jones had put the blue bottle, and when it was next seen it had been washed out. The doctor, however, scooped up some crystals from the carpet, which were shown to be strychnine hydrochloride.

Bland testified that Vaquier had visited him on March 1st, and this time had asked not only for the usual cobalt nitrate, copper acetate and so on, but for strychnine, which he needed for his radio experiments. When the chemist demurred, Vaquier told him that he could get "any amount he needed" in France.[13] Faced with this illogical argument, Bland gave in and deferred to the other's expertise as a wireless experimenter. He sold him 120 milligrams of the hydrochloride (long crystals), for which Vaquier signed the poisons register in the name of J. Wanker. (Bland testified, "They spell their names so curiously, some of these foreigners. I was not surprised.") The strychnine cost twopence. An employee of the Marconi company was called to testify that strychnine played no part in wireless telegraphy.

For some time since, the accused could give evidence on his own behalf in a criminal trial, but Vaquier, who had been a popular character during his stay in Surrey, proved his own worst enemy. Before being charged, he gave a flamboyant newspaper interview, as a result of which Bland recognised his photograph. Not only did Bland pick out Vaquier at an identity parade, but Vaquier stepped forward to meet him in the police station as if greeting an old friend. Worse was to come during the trial, for Vaquier treated the whole occasion as an excuse to display his prodigious ego, combing his beard and brilliantining his hair in the witness-box. Whether, as some have surmised, he was mislead by the low-key progress of the English legal proceedings into thinking that things were going well for him, is a matter for speculation. If he thought that he would be able to stage manage a courtroom *dénouement* justifying a *crime passionnel*, he was mistaken. Certainly, when found guilty he let loose a torrent of abuse at the judge and jury.

With the help of interpreters, he made no less than five written statements to the police before he had even been arrested. These variously accused the Joneses of keeping a brothel; denied any intimate relationship with Mabel Jones; accused two different men of being her lover, and one of them (unidentified) of having been the poisoner; described how he and the Joneses were on the point of moving to France to open a hotel; and explained that had he wanted to poison Jones, it would have been much easier for him to have done it on the Friday night when Jones was dead drunk. Cross-examined in court, he contradicted himself with abandon, and revealed for the first time the name of the man whom he accused: Mrs. Jones's solicitor, Bruce Millar.

Vaquier's accusations against Millar were ludicrous. He described how this man, to whom he had never before spoken and whose name he did not know, buttonholed him at the Blue Anchor one Sunday and asked him if he knew where he could get some strychnine to kill a dog with the mange, since although he worked in London, he did not have time to buy it himself. Millar gave him one pound, enough to buy a sack of strychnine, and never asked for change. As a result, Vaquier went and bought 25 grams (not 120 milligrams) of strychnine, and signed a false name in the poisons register.

Q: What did you think you had been asked to sign your name for?

A: I attached no importance to it.

Q: If that is so, why did you not put your real name?

A: Because I had been told that when you buy poison, you never sign your own name.

Q: Who told you that?

A: The solicitor…

It was hardly necessary for Millar himself to be cross-examined to rebut these allegations.

Analytical methods now in use were still based on colour reactions like those of Taylor's era, but refined and made more accurate. The depth of colour produced by the chemical reaction of the unknown strychnine sample would be compared with that of a similar solution of known strychnine strength using a simple optical instrument. This method had been invented by Duboscq at the time of the Palmer trial, but did not catch on until about 1870 when it became the standard analytical method until about 1950.[12] The way that the results were reported in court was still anachronistic, however. In the Vaquier case, John Webster, home office analyst, said that he had detected a total of "a little over seventeen thirtieths of a grain" of strychnine in the various organs of the deceased. When pressed, he said that it did not amount to as much as eighteen thirtieths.

There was a discrepancy in the strychnine quantities. The only purchase that was ever definitely traced to Vaquier was the 120 milligrams (less than two grains). This was enough to kill two or three times over, certainly, but surely nowhere near enough to ensure that a man taking a spoonful out of a jar leaving half behind, spitting and vomiting much of it out and spilling it on the carpet, could end up with the 37 milligrams detectable in the body that Webster found. There was much discussion on this point, with the forensic scientist Bernard Spilsbury being cross-examined about the minutiae of how caking of the bromo salts as a result of exposure to the atmosphere might concentrate the strychnine into a zone of loose powder, and so on. But the impression remained that the several grams in the bottom of the blue bottle must have been nearly all strychnine hydrochloride. This might just conceivably have been grounds for an appeal had Vaquier kept his mouth shut again. But while in prison he made a statement that after the death he had seen a woman—Mrs. Jones or someone else—go to the toolshed at the Blue Anchor, and that he had afterwards seen behind a loose brick a bottle with strychnine in it. The police found the bottle in the place indicated. It contained 23 grains (nearly one and a half grams) of strychnine hydrochloride. There can be no doubt that it belonged to Vaquier himself, although the purchase was never traced. Bland may have sold him no less than 25 grams (385 grains, enough to kill 700 people), but the truth was never established.

The skills of Robert Robinson represented the culmination of the development of "classical" organic chemistry, which was based on chemical skills alone; carrying out chemical reactions on the unknown substance, then using pure logic to interpret the results. The achievements of Robinson and many others using these immensely time-consuming methods were stupendous, but exceedingly labour intensive. By the latter part of Robinson's career, a new kind of chemistry was emerging, one based on the use of instrumental methods to aid the work of the experimenter.

Robert Burns Woodward is thought by many to have been the greatest of all organic chemists, although it is impossible to make objective comparisons with those born in other eras. His date of birth (1917) certainly positioned him chronologically to capitalise on the possibilities of the new technologies, but he had the brilliance to exploit them to the full, and played the leading role in showing his contemporaries how to.

Woodward's father died in the influenza epidemic of 1918. Left a widow, Margaret Woodward remarried but was soon abandoned by her second husband, and brought her son up alone. Robert was fiercely proud of his Scottish roots and the fact that his mother, whose maiden name was Burns, had named him Robert after the poet, though he did not go so far as to claim direct descent.

The only child working alone in his basement laboratory is a Hollywood cliché, but Woodward enacted this role to perfection and carried it to extremes. By the age of 12 he had carried out every experiment in Gattermann's *Practical Methods of Organic Chemistry,* then the standard text for college students. The next few years reinforce the stereotype of the difficult genius. He went to the Massachusetts Institute of Technology at the age of only 16, and was found to know more chemistry than any of the graduates; after only a year, MIT took the extraordinary step of giving him his own laboratory. But he skipped chemistry classes that he thought simplistic or irrelevant, and refused to fulfill the mandatory nonchemical course requirements, such as physical education. As a result he was suspended for a period and only re-enrolled after the professor of organic chemistry, James F. Norris, took a personal hand. As Norris told a reporter from the *Boston Globe* in 1937:

> We saw we had a person who possessed a very unusual mind in our midst. We wanted to let it function at its best. If red tape which was necessary for other less brilliant students had to go we cut it. We did for Woodward what we would have done for no other student in our Department, for we had no student like him in the Department. And we think he will make a name for himself in the scientific world.

As with Robinson, even the *Festschrift* publications written much later to celebrate Woodward's achievements do not always gloss over his demanding personality. One, published in 1992 well after his death, and otherwise a meticulous report of his positive accomplishments, says:

> After receiving his Ph.D. Woodward taught summer school in the chemistry department at the University of Illinois, where he managed through his intelligence and impatience to alienate several of the leading figures in organic chemistry in the United States.

Like Robinson, Woodward burst upon the scene with a stupendous synthetic achievement while still in his twenties. In 1944 he published the total synthesis of quinine, and followed this with syntheses of cholesterol and cortisone. By 1950 he was full professor running a large research group at Harvard, where his flamboyant style, fuelled by coffee, cigarettes and Scotch whisky, became famous. His lectures, invariably given in his trademark light blue suits, lasted for several hours, with meticulous blackboard schemes drawn in several different coloured chalks.

Peter Bladon, who worked at Harvard for a year towards the end of the decade, remembers the atmosphere of Woodward's research group.

> I was well used to working late; most chemistry postgraduate students were, and are; it's a highly labour-intensive activity. But Woodward's Thursday seminars were something else. They started at 8 p.m. and went on until the small hours. The invited speaker would be listened to in silence while everyone

else prepared their contributions. As soon as he sat down they would all spring to their feet. Sometimes there would be three or four people drawing on the blackboard simultaneously. Then when they had all finished for a bit, Woodward would get up and meticulously demolish everything that had gone before, including the guest speaker. He was a very cold, single-minded individual. It was dog eat dog. If you didn't come up to scratch in terms of your intellect or more especially your results, he ignored you, and you would just drift away.[15]

Woodward's personal life was something of a closed book; superficially conventional, but not a subject for close scrutiny. The record books report a first marriage in 1938 to a fellow-student of Finnish descent, Irja Pullman, a second to Eudoxia Muller about eight years later, and two children by each wife. The caption to a 1977 photograph reads:

The arrival of the telegram announcing that Woodward had won the Nobel Prize for the totality of his work in chemistry. Dolores ("Dodie") Dyer, RBW's loyal secretary and friend, looks on.

In August 1954, Woodward and his team astonished the scientific world by publishing a two-page note in the *Journal of the American Chemical Society* claiming the total synthesis of the strychnine molecule, starting with a simple laboratory chemical that contained only one ring of the final structure. This structure of strychnine had only just been finally settled, and virtually every other chemist would have thought that assembling its seven-ring structure in exactly the right way an outrageous impossibility for the methods of the time.

Organic chemical reactions do not usually go in 100% yield. In other words, for every 100 molecules of the starting compound, only say 80 will be converted during the reaction to the desired end product. The other 20 will go off on various other pathways to end up as undesired byproducts. Not only does this result in the loss of 20% of the material, but the byproducts may well interfere with the process of purification, thus effectively reducing the yield even further. The problem becomes acute when the synthesis is a multistage one. A simple calculation will show that after a 10-stage sequence in which each step gives 80% yield, the overall yield of the sequence will be only 10.7%. The skill lies in designing the conditions for each stage to produce precisely the required molecule in the highest possible yield. The Woodward synthesis of strychnine required 28 stages. There must have been some who doubted its veracity or reliability. Woodward was not fazed by the doubters; he just went on to more and more peaks of achievement. It is true, however that the full details of his strychnine synthesis were not published until 1963, when it could be seen that the yield is in fact minuscule.[16] The publication of the bare details of the synthesis without giving the yields until nearly a decade later was a tour de force on Woodward's part, silencing any potential cavillers who might have objected that a synthesis in such minute yield was hardly a synthesis at all. Woodward need not have worried. The deliberate design and execution of the synthesis of such a complex molecule was an unprecedented achievement, and together with his other syntheses, it lifted the science of chemistry to a new level. His remained the only synthesis of strychnine for nearly forty years, and he won the Nobel prize in 1965 for his achievements in synthesis.[17]

This synthesis of strychnine had no direct practical utility, for the price of synthetic strychnine was, and remains, millions of times higher than the natural product. But by pushing back the frontiers of what could be done, Woodward, who died in 1979 at the age of only 62, developed science to a new level, with incalculable repercussions.

The new techniques of chemistry transformed not only laboratory synthetic chemistry, but with even greater impact if that were possible, analytical chemistry, of which forensic analysis is a subdiscipline. The colour test for strychnine is a thing of the past. A variety of machine and other techniques are now available to detect minute traces of multiple drugs in the blood and tissues of a live athlete or a dead body.

Thin-layer chromatography (TLC) is relatively insensitive, but can be used to carry out a rough screening for several hundred substances at the same time, at very low cost. Radioimmunoassay costs a little more and can detect one billionth of a gram, and mass spectrometry, which is more elaborate, can detect and identify as little as 10^{-12} grams, or a thousandth of a billionth of a gram (a picogram). There's no way to escape the infallible eye of the forces of scientific law and order, unless you are able to persuade them to look in some other direction altogether.

Endnotes

1. Bernays, 1855; 388.
2. Williams, 1990.
3. Todd and Cornforth, 1976:415.
4. The passage on W.N. Haworth in Robinson's autobiography is a masterpiece of backbiting. He ignores all Haworth's greatest attainments, such as determining the structure of vitamin C and winning the Nobel prize, describes him as a researcher on linoleum, accuses him of stealing all of his achievements from E.L. Hirst, misspells his name as "Howarth" and transposes the photographs of Haworth and Hirst. This was written more than 25 years after Haworth's death.
5. The Clarendon chemical laboratory in Oxford is even today far more difficult for a visitor to gain entry to than other comparable departments such as the equivalent Lensfield Road building in Cambridge. This may be the result of a culture dating back to Robinson's suspicious nature, and cannot be totally unrelated to the fact that for many years, the chemists there had been handling poisons on such a gigantic scale.
6. Robinson, 1952.
7. The Robinson group published more than 250 papers on strychnine.
8. Robinson, 1952, loc. cit.
9. Robinson, 1976.
10. Williams, 1990.
11. Smyth, 1980.
12. Blundell and Seaton, 1929.
13. Not true! See Chapter 7.
14. Rosenfeld, 1999:255.
15. Prof. Peter Bladon, personal communication.
16. The overall yield of the Woodward strychnine synthesis, not given in his paper, can be calculated by multiplying together the yields reported for the individual steps. It is 0.00006%, which means that to prepare one gram (less than half a teaspoonful) of pure strychnine would necessitate starting with 1,600 kilograms of the starting material. Even this figure is an underestimate, because in several of the reactions, the yield stated is for crude material which needed further purification, with inevitable losses, before it could be used in the next step.
17. He would have shared in the 1981 prize too had he not died prematurely. It should, however, be mentioned that recent work has thrown some doubt on his 1944 quinine synthesis.

CHAPTER **16**
I Didn't Know It Was
Used for Poisoning

The quickest way of ending a war is to lose it.

George Orwell, *Second Thoughts
on James Burnham*, 1946

The opening days of 1917 marked the nadir of Allied fortunes in the First World War. John Grigg[1] argues persuasively that Britain's situation at the time was even more perilous than that faced during the Battle of Britain in 1940. On the Western Front there was stalemate. Relations with a devastated France were under strain, and morale at home was affected by hundreds of thousands of casualties, with no end in sight. The United States would not enter the war until April, and as yet it was not certain that this would happen. The Eastern Front was crumbling away, and within weeks the February revolution would lead to the abdication of the Tsar, the separate peace between Russia and Germany, and the consequent diversion of many divisions of German troops to the West. Most perilous of all for Britain was the submarine threat to the shipping lanes and the importation of the goods from abroad needed to keep the war going.

As in 1940, the situation demanded a gifted and charismatic leader. This was David Lloyd George. At almost the same instant that he had become prime minister in December 1916, the German high command had switched their priorities to make destruction of the British merchant navy their main aim, and had again declared unrestricted U-boat warfare, a programme which was very nearly successful. Among his many other achievements, it was Lloyd George whose organisational and political skills nationalised shipping, created the Ministry of Shipping staffed by some exceptionally talented civilians, and steamrollered the admiralty into accepting the convoy system. Had he not existed, or had been taken away at this crucial juncture, the Allies might well have lost the war.

And yet this nearly happened, for in those opening weeks of 1917 there was a plot, now virtually forgotten, to assassinate Lloyd George and Arthur Henderson, leader of the Labour Party and member of the war Cabinet, using strychnine.[2]

It is not necessary to dwell any further on the question of how serious the consequences of such an assassination would have been; they would have been catastrophic. But how real was the threat, and how likely was it to succeed?

The accused were all members of the same family. They were a middle-aged woman, Alice Wheeldon, a dealer in second-hand clothes, her two daughters, Hettie Wheeldon and Winnie Mason, and Winnie's husband, Alfred Mason. Hettie lived with her mother at 12 Pear-tree Road in Derby; the Masons lived in Southampton where Alfred was a lecturer in pharmacy at the University College. Apart from Alfred's access to strychnine as part of his profession, superficially they seem to have been as mundane a lower-middle-class family as one could have wished to find. At the first hearing at Derby assizes, there were echoes of the Christiana Edmunds trial of the 1870s in the way the court reporters, bemused by goings-on beyond their ken, sought reassurance in falling back on observation of the appearance of the people in the dock, and all the usual hackneyed prejudices received another airing. "The woman Winnie Mason," the Weekly Dispatch noted,[3] "Who is dressed in a red cloak and wears a soft, black velvet hat adorned with a coquettish bow, turns an attractive face to her husband." Hettie Wheeldon, the paper opined, was of the two the less attractive, but there was no mistaking the fact that they were sisters. She looked like a schoolteacher, which she was, and like Christiana Edmunds nearly forty years before, she took copious notes in court.

The attitude of all four prisoners was one of unconcern. This was especially marked in the case of the male prisoner, called by his mother-in-law "Alf." "He has an abundance of fair hair, which he arranges after the manner of the intellectuals," the paper went on, "Many socialist voters look just like this."

Inspector Edward Parker of Scotland Yard laid the charge: "That on diverse dates between 26th December 1916 and 29th January 1917 they did amongst themselves unlawfully and wickedly conspire, confederate and agree together, one the Rt. Honourable David Lloyd George and one the Rt. Honourable Arthur Henderson, wilfully of their malice aforethought to kill and murder contrary to the Offences Against the Person Act 1861, section 4, and against the peace of our Lord the King, his crown and dignity."

The hearings were a long, drawn-out affair. The prosecution case was first of all made out at Derby, the prisoners reserving their defence, then again in full at the Old Bailey in London the following month. During the latter trial one of the jurors became ill, so that the trial had to be restarted. The newspaper reports give the impression that by the time that the whole of the prosecution evidence had been gone through three times, it began to seem both preposterous and banal. But there were some bizarre twists when the defence case was eventually presented.

Mrs. Wheeldon, driven by sardonic rage, was the motivating force of this strange family; according to the prosecution witnesses, she sounded off like a fishwife at home, using "extremely obscene and blasphemous language" whenever Lloyd George's name came up. She said that he had been the cause of millions of lives lost, and that "George in Buckingham Palace" should also be "done in." Extracts from letters penned by her daughters and read in court demonstrate that they were hardly her inferiors in abuse: "Who can blame the Germans now for taking their revenge on rhinoceros-skinned, perfidious canting Britain," Hettie wrote, while Winnie to her mother, "You know what slimy cowards the pious Christian English are if their dirty skins are in danger. The only ones to trust are ourselves."

These remarks immediately raise the question of whether there was an Irish connection. The First World War was a desperately fraught time in Anglo-Irish relations. Far from sweeping this possibility under the carpet, the prosecution probed it. At the Old

Bailey, Hettie Wheeldon was asked if she was English. She told the court, "I was born here and so were my Parents as far as I know."[4] Since no barrister ever asks a question to which he or she does not already know the answer, this question was almost certainly inserted at the behest of the authorities in order to defuse any rumours that might have been circulating. She was not specifically asked if the family had any Irish connections. According to the biography of F.E. Smith (Lord Birkenhead), the attorney general who led the prosecution, neither was any German connection found.[5]

Mrs. Wheeldon was the centre of a small network of people helping conscientious objectors hide from the authorities, and it is clear that it was this activity that led to them taking an interest in her. When the police went to 12 Pear-tree Road, they found and arrested Alexander McDonald, a deserter from the Sherwood Forester regiment. On being charged, Mrs. Wheeldon said, "I think it is a trumped-up charge to punish me for my lad being a conscientious objector. You punished me through him while you had him in prison last time."[6] During the trial, her son, William Wheeldon, was arrested again in Southampton and handed over to the military authorities, but escaped.[7] The son-in-law Alfred Mason thought that as a trained chemist he was exempt from military service, although he did not have a certificate of exemption; he was not trying to evade service, though he would appeal if he were called up, he said; he was not a fugitive. Winnie Mason sardonically testified that she had been used to hearing her mother use violent expressions in referring to Lloyd George, but "attached no more importance to them than she did to similar remarks made by people who expressed a desire to hang, draw and quarter conscientious objectors."[8]

It is difficult now for us to get inside the minds of such people at such a time. The country was fighting for its life and it was a period during which deserters from the front were regularly executed. Secret agents roamed the streets of Britain hunting them down, like the shadowy Major Lee, who ran the department responsible for introducing Messrs. Gordon and Booth into the Wheeldon household as agents provocateurs. Gordon introduced Booth to the Wheeldons as "comrade Birt of the Independent World Workers." Mrs. Wheeldon told Booth, "We had a plan before when we suffragettes spent £300 in trying to poison him [Lloyd George]. It was to get a poison in a hotel where he stayed and drive a nail through his boot which had been dipped in poison." Booth asked how the poison was used. She said, "It is a crystal, and you drop two drops of water on it, dip your article in and when it evaporates it leaves the poison."[9] She also claimed responsibility, more or less, for having set fire to Breadsall Church near Derby.[10] If this was true, which it probably was, it weakens her status as a war protester and leaves her as an obsessive crackpot, for the church was burnt in June 1914, before the war started.

Amongst the scattered legal papers, the chief object of interest on the court table at the trial was a tobacco tin containing four phials, apparently purloined from Guy's Hospital in London by Alfred Mason and sent by him from Southampton to Derby together with instructions for use. Phial A contained 7½ grams of crystalline strychnine hydrochloride, labelled by Alf as being "Sufficient for two or even three doses to be given by the mouth or in solution." (The prosecution stated that this was "enough to kill 500 people;" in fact the true figure is about 150.) Phial B contained some of the same in solution in water. Phial C contained a powder which was shown to be curare: "To be injected either in solution or by a dart, which will penetrate into the body and stop for a while. Rusted in solution or fired from an air-gun, or a rusty needle if driven well in with powder will do, but don't advise unless in urgent dilemma." Phial D was the same in solution.

Curare is a very powerful poison related to strychnine but is inactive by mouth. Strychnine could be given by mouth or by injection. The rusty-nail or air-gun injection technique would not have worked with strychnine since the amount that could be introduced into the body by this technique would not have been sufficient. It might have worked with curare, as it did with ricin some six decades later, when Georgi Markov was assassinated on Waterloo Bridge, London in 1978 by a KGB agent using a poisoned umbrella. But the general tone of the instructions from Alf Mason, who was clearly a disorganised fantasist, combined with the inside story of life at the Wheeldons' house in Derby, as revealed by the secret agents, casts doubt on whether the plot would ever have materialised in a realistic way. The fact that Mrs. Wheeldon and her associates, whom she claimed had spent £300 on setting up the earlier plot to kill Lloyd George, had become discouraged when he "went to France"—hardly an unexpected development or an insuperable hurdle—speaks volumes for their lack of true determination.

The defence case, when it finally surfaced at the Old Bailey, was a timeworn one. The strychnine and curare were needed to poison dogs used to guard the camps in which conscientious objectors were held. Incidentally, Bernard Spilsbury, the famous forensic scientist, in giving evidence, stated that dogs could not taste strychnine.[11] Alfred Mason, cross-examined, said, highly implausibly, that although he had poisoned two thousand dogs with strychnine (likewise incredible), he had no idea that it had been used for the purpose of murdering humans.[12]

The defence counsel, Mr. Riza, blustered that, "This is a scandalous, a vile, a vindictive prosecution and most dangerous, the first of its kind in England." Who precisely, he demanded to know, were these secret agents, who had not given evidence? He went on to make the fatuous suggestion that if Gordon were not to be cross-examined, the only course left open to him was to offer his clients for trial by ordeal.

Mr. Justice Low: I'm afraid that has been abolished.

Mr. Riza: That is why I submit it to the jury.

Mr. Justice Low: That the ladies should walk over hot ploughshares or something of that kind? Is that it?

Riza must have been aware that many among the pro-war majority in the country would have had the Wheeldons shot, and that if they escaped the maximum sentence of ten years for the conspiracy, they would face other charges connected with aiding the escape of enlisted men, which could be equally dangerous. In the light of this general climate, it says much for the British that as the trial unfolded and the plot found to be amateurish, there was scope for these few jokes in court. And it says much for the character of Lloyd George himself that before the end of 1917, Mrs. Wheeldon, ill in Holloway prison as a result of a hunger strike, was released on his express wishes.[13]

To say that the trials were literally a nine-day wonder would hardly be overstating the case. So much desperation was vying for space in the papers, that by the time the full trial had taken place at the Old Bailey and sentence pronounced, the affair had already become stale news. The *Weekly Dispatch* in its issue of February 14 1917 devoted four columns to the hearing at Derby. Five weeks later, it could hardly stir itself to announce, in an item baldly headed "Latest News" and running for a mere two column inches, the sentences: Mrs. Wheeldon, ten years (the maximum sentence); Alfred Mason, seven years; Winnie Mason, five years; Hettie Wheeldon, not guilty. Since the paper was published on Sundays, there would have been ample time to prepare a lengthy retrospective, but it appears the editor thought that the public had lost interest. Or did its owner, Lord Northcliffe, ensure that after the initial lurid poisoning sensation, no space was available for any possibly morale-sapping post-mortem as to why four citizens of Middle England had apparently felt such rage against their country? The *Illustrated London News* similarly gave a page, with photographs, to the Derby hearing, but when the eventual verdicts were announced they only generated an incongruous article in its "Science Jottings" column about poisoned arrows and their use in the world's jungles. Even then, the author could not avoid the temptation to finish with a reference to the obsessive topic of the day: German beastliness. The primitive peoples of the world could not be blamed for using such a deadly substance because they used it to get their food, but "its use today among 'Civilised' peoples is now confined to criminals—and to apostles of 'Kultur'."[14]

At the conclusion of the trial, the suffragette leader Mrs. Emily Pankhurst, who had been in court earlier in the hearings, asked to address the court from the witness box. She disowned Mrs. Wheeldon's statement about the suffragette movement having spent money on a plot to kill Lloyd George, and said that her organisation regarded the prime minister's life as of the greatest value in the current grave crisis. This was no less than the truth, for Lloyd George did indeed pilot women's suffrage through parliament as soon as peace arrived. "The Man Who Won The War" lived another twenty-eight years and nearly saw another war out, dying in his bed on March 26, 1945.

Endnotes

1. Grigg, 2002.
2. *Derby Daily Express*, 31 Jan. 1917:4; 1 Feb. 1917:1; 3 Feb. 1917:1; 5 Feb., 1917:1; 6 Feb. 1917:4.
3. *Weekly Dispatch*, 14.2.1917:1–2. This was a national weekly news-paper, later the *Sunday Dispatch*. It closed in 1961. In 1917 it was owned by Lord Northcliffe (Alfred Harmsworth) who also owned the *Times* and the *Daily Mail*. Within a few weeks of the Wheeldon trials he was to play an important unofficial role for Lloyd George's government, touring the United States to coordinate the war effort with the new ally.
4. *Times*, 8 March 1917.
5. Birkenhead, 1935:2:80–84.
6. *Times*, 1 Feb. 1917:7.
7. *Times*, 7 Feb. 1917:3.
8. *Times*, 8 March 1917.
9. *Times*, 4 Feb. 1917:4.
10. Suffragettes were blamed for the fire, which destroyed many valuable books, but their involvement was never proved. http://www.breadsallparishchurch.co.uk/index.html.
11. *Times*, 8 March 1917.
12. Ibid. Thompson (1925): 192 incorrectly says that Mason said "he did not know that strychnine was used for poisoning," which would have been even more preposterous. Mason missed the golden opportunity to cite da Orta in his defence to prove that it was only poisonous to dogs, not humans.
13. *Times*, 31 Dec. 1917:3.
14. *Illustrated London News*, Feb. 10 1917:174; March 24 1917:342.

CHAPTER 17

Is There a Faceless Fiend?

"I do not know," said Poirot. "But one thing does strike me. No doubt it has struck you too."

"What is that?"

"Why, that there is altogether too much strychnine about this case. This is the third time we have run up against it."

Agatha Christie, *The Mysterious Affair at Styles*, 1920

By the end of the nineteenth century, the century that had given strychnine birth, its deadly properties, together with its periodic courtroom appearances in some of the most lurid of all poisoning cases, had ensured that it had entered mythology.

The process began with *The Count of Monte Cristo*, published in 1844, which features both strychnine and brucine. The two poisons were heaven-sent to play a central role in this picaresque farrago of double-dealing, revenge, suicide by poison and murder by poison, not to mention narrow escapes from murder by poison, poisonings in mistake for someone else, resurrection after presumed poisoning, and burials alive. There is a pleasing intertwining of events in the real and fictional worlds. As (both in real life and in the book) Louis XVIII is restored to the French throne following the battle of Waterloo, in the real world Fouquier is communicating the results of his experiments on paralytic patients at the Hôpital de la Charité, while in the dungeons of the fictional Château d'If at the other end of France, the Abbé Faria is suffering a seizure which leaves him paralysed in an arm and a leg, and is being treated with a phial of medicine that he has secreted in the leg of his bed. Later on there are realistic descriptions of the expiries of Madame de Saint-Meran and of the servant Barrois from brucine poisoning, the latter after drinking brucine-containing lemonade intended for M. Noirtier, but there is also a great deal of mystical rambling about the rituals of poisoning, owing more to the Arab writers of the Middle Ages than to the nineteenth century. Indeed when the subject of poisons first comes up in the book, Dumas has the poisoner, Madame de Villefort, explain her unexpected knowledge of the plant from which brucine comes by the statement, "I have a passion for the occult sciences."

As a young man, Dumas followed the 1823 trial of Dr. Castaing with great interest. Barrois's symptoms in *The Count of Monte Cristo* closely resemble those of Auguste Ballet as described in the trial reports and as Dumas himself recorded in the pages of his journals, and the lemonade is too much of a similarity to be coincidental.

Dumas perpetrates two misconceptions. The first is that brucine is an habituating poison, so that someone could immunise himself against its effects by taking one milligram the first day, two the second and so on, until after a month, he would be able to share a carafe of poisoned water with the intended victim. According to the plot, the reason that Noirtier, who also drinks some of the lemonade, does not succumb is because he is completely paralysed, and his doctor has for some time been treating him *au Fouquier* with brucine to cure it. Dumas also perpetuates the myth that strychnine is so poisonous that a man who ate a fowl that had pecked at the body of a rabbit that had eaten a cabbage that had been watered with it, would die. Apart from anything else, strychnine is strongly adsorbed by soil, and is not taken up by growing plants. Had Dumas written his book a few years later, he might have based it more on the true revelations emanating from Rugeley than on the arcane beliefs of some of the French physicians of the Fouquier era.

Liebig, the famous chemist, despaired. At the time of the Burton Ale scare in 1851–1852, there was public hysteria about the possibility of being poisoned by strychnine in beer, which Liebig attributed to the man in the street's knowledge of the alkaloid being based entirely on Dumas's fantasies and not on the scientific facts.

Exactly as happens with many human celebrities, strychnine's literary fame really took off, later in the century, just as its power in the real world began to decline. Helen Mathers (Mrs. Henry Reeves) was an immensely popular but now forgotten novelist of the fin-de-siècle overblown gothic school, full of pale heroines with names like Nadège Straubenzee. Her offering *The Sin of Hagar* introduces a black villain, Blake Trelawney, and an insane but beautiful villainess, Hagar Gregorius, brought up as the daughter of a "cadaverous, cold" scientist who sacrificed her on "the wicked altar of man's thirst for discovery" in experiments involving mesmerism, chemistry and paedophile sessions. Hagar tries to hypnotise the heroine into poisoning her husband. A bystander, Lady Lirriper, detects this immediately because the strychnine discoloured a whisky and soda. (It doesn't.) Because she is "a good amateur chemist," within a few seconds of snatching the poison out of the heroine's hand, she had "carefully examined the contents of the tumbler" and confirmed the identity of the poison. (Where was she when the Palmer trial needed her?)

There is no indication that the narrative is set in any period but the present, that is 1896, but the observation that news travels slowly is supported by the appearance of a former associate of Ms. Gregorius's father, a Frenchman who tries to save her from the abyss. This leads the author to speculate "was her father one of the salpêtrière group, the results of whose experiments burst some years ago upon an astonished world?" The year 1896 was as far away in time from the discovery of strychnine as is the present day from John Logie Baird's television experiments.

Mathers's more distinguished contemporary was Sir Arthur Conan Doyle, and strychnine merits a number of passing mentions in his stories. But, as many others have pointed out, although Sherlock Holmes is said in the very first of them to have a "profound" knowledge of chemistry, the part that this, or indeed any other technical knowledge, plays in the successful resolution of his mysteries is vanishingly small.

A more intricate web, and one making unique connections between the literary and the scientific, was spun by another forgotten novelist, John Uri Lloyd, in his peculiar novel *Stringtown on the Pike, A Tale of Northernmost Kentucky*, written in 1900 but set in the period immediately after the Civil War. The first half of the book is virtually irrelevant; it is full of dialogue involving former plantation slaves written in a vernacular which is meant to be picturesque but now comes over as more or less racist as well as highly tedious. The real story begins after the midway point. The narrator, Samuel Drew, leaves Stringtown after an unhappy love affair, vowing never to return, and becomes professor of chemistry at a northern university.

Years later he receives a request to carry out the forensic examination of the stomach of a Stringtown man suspected of having been poisoned. He tests for strychnine with positive results: "I was fairly well satisfied, although there were some points in connection with the chrome-sulphuric acid test which puzzled me. The blue-violet color surely did appear, but it was not as characteristic as I would have liked." As he ponders the result, he receives a surprise visit from his former paramour, Susie, who discloses that the man accused of the poisoning is none other than his rival for her hand. Ignoring her entreaties, and her attempts to bribe him with a chestful of gold surfacing like an article of *Titanic* salvage from an earlier sunken part of the plot, he resolves, after again repeating the tests, to follow the path of true science. He dispassionately suppresses his loathing for the accused, and returns to Stringtown against all advice to give evidence; not only does he tell the jury what he found, but he demonstrates the colour test to them in the courtroom.

The judge is probing: "Have you tried the test with every plant, shrub, tree, leaf, root, bark, fruit, that grows?" he demands. The defence attorney is blustering: "That man comes here from the North, he touches a liquid with a bit of stuff, and it turns blue, violet blue, for an instant. He asks you, men of Stringtown County, to hang a resident of Stringtown County, because this blue color comes in a dish. Kentuckians, if you become a party to this monstrous crime, a few dollars hereafter will lure a horde of hungry chemists from the North to show a color in a dish to whoever cares wrongly to gain an inheritance or wishes to hang an enemy....these ghouls will find chemistry to show that all who die are poisoned.[1] There sits a man who comes back to us to hang the companion of his youth. One of them is a *man*, the other a *chemist*!" The jury convicts, but for reasons which need not detain us here, the accused is killed in a courtroom shootout before he can be sentenced. The chemist goes back to his laboratory.

Drew then receives a visit from Susie, who tells him that she will marry him once she is satisfied that the verdict was a just one. In order to do this, she proposes to follow a course of study in chemistry.... The two work away at the bench together for some considerable chaste period, then she departs for further studies elsewhere. On her return she reveals that she has proved that the characteristic blue colour test for strychnine with dichromate-sulphuric acid is also given by a mixture of the alkaloids hydrastine and morphine. Hydrastine is contained in the plant goldenseal (*Hydrastis canadensis*), which the dead man used to take in a tonic, and morphine had been administered to him by the doctors when he began to fall ill. Game set and match to the member of the weaker sex, who takes out a vow of eternal celibacy, and shame, nothing but shame to the chemist, who slowly poisons himself to death with hydrastine, in a dénouement almost uniquely depressing.

Lloyd, the author, was a well-known herbal pharmacist and founder of the Lloyd Library in Cincinnati, the world's largest collection of herbal writings. It appears that in a bizarre retooling of the normal process of scientific publication and peer review, he had discovered the abnormal colour reaction of hydrastine/morphine himself, and had put it into his novel rather than publish it in a scientific journal. Because he was such a reputable scientist, the book, which was widely read, was subsequently cited as established evidence in a number of American criminal trials to undermine the credibility of the prosecution's forensic evidence on the colour test for strychnine. It seems unlikely that there were ever any genuine cases of goldenseal interference. But the book passed into mythology so effectively that even today, in an era of virtually infallible chromatography and mass spectrometry, there are still some athletes who believe that taking goldenseal will prevent the detection of every known drug.[2]

An even larger proportion of the ongoing mythology concerning the very name of strychnine can be laid at the door of Agatha Christie (1890–1976). This seems to be a woman obsessed with drugs and poisons; she served as a dispenser during both world wars. Nearly every novel, and many of her short stories, contains some reference to people dosing themselves, or other people.[3] *A Caribbean Mystery* (1964) mentions no less than 22 medications, ranging from linseed poultices to face cream enhanced with belladonna.

In the novel with which she burst upon the scene, *The Mysterious Affair at Styles* (1920), the lady of the house, Mrs. Inglethorp, is poisoned by strychnine while in a locked bedroom in the middle of the night. The plotting already carries all of Christie's trademarks of (over)ingenuity and countless red herrings.

Christie had clearly done her strychnine homework. In *The Mysterious Affair at Styles,* there are numerous echoes of events earlier in this book. For example:

- A coffee pot left unattended in a hallway where it could have been dosed with strychnine. Servants unknowingly taking poisoned dishes up to a bedroom. The victim dying in the small hours (Palmer).
- A doctor who is suspiciously up and dressed in the middle of the night (Palmer).
- A chemist's assistant testifying that he sold strychnine to someone who claimed he needed it to kill a dog (Palmer, Dove).
- An inquest at a village inn (Palmer).
- A pharmacist running frantically up the drive fearful that someone has been poisoned by strychnine (Mrs. Sergison Smith).

When it is recalled that the latter case was fully described in court during the Palmer trial, it is clear that Christie must have based her research very substantially on events at Rugeley and the Old Bailey sixty years earlier.

But the forensic part of the plotting is likely to leave the reader feeling cheated.

The technique employed for the successful poisoning depends on the fact that a solution containing strychnine sulphate and potassium bromide precipitates an insoluble salt on standing, so that nearly all the strychnine ends up in the bottom of the bottle. The murderers add bromide to Mrs. Inglethorpe's medicine, then wait around for her to reach the last dose. We are expected to believe that the medically unqualified conspirators, whose main talent appears to be running around the village wearing false beards, are party to an obscure piece of physical chemistry that only Christie, a qualified dispenser, and at most a handful of her readers, would have known.[4] The *Pharmaceutical Journal*, however, thought the book "first-rate," although it rather patronisingly thought that the female author either had a pharmaceutical qualification or "had called in a capable pharmacist to help her over the technical part."[5]

The detective Hercule Poirot makes his first appearance in the book. Several possible role models for her Belgian hero have been floated, for example G.K. Chesterton's fictional Hercule Flambeau.[6] When questioned much later, the somewhat uncommunicative Christie said that Poirot was not based on any one person, factual or fictional, but that her choice came from observing Belgian refugees in her native Torquay during the First World War. But given Christie's extensive reading on forensic detection and the history of alkaloid detection, a much more attractive candidate is the real-life Belgian chemist Jean Servais Stas, whose successful detection of nicotine at the Château de Bitremont in the 1840s, as described in Chapter 7, led to the execution of Count Hippolyte de Bocarmé and the foundation of modern forensic science.

Strychnine makes three further appearances as murder weapon in Christie's work in: *The Coming of Mr. Quin* (1930), *Death on the Nile* (a short story) (1934) and *How Does Your Garden Grow* (1935). It is the tool of unsuccessful attempts in *The Adventure of the Egyptian Tomb* (1924), where Poirot catches a murderer by pretending to have been poisoned by something horrible and convulsive, and *The Blue Geranium* (1932), in which a patient declines an injection later found to be strychnine. In *The Big Four* (1927) and *A Pocket Full of Rye* (1953) it is suspected but disproved, and in two further books it gets passing mention.[7] In none of these later stories does the chemistry of strychnine play a central part; Christie must have felt that she had more or less worked this particular seam out, and her attitude on the whole is summarised by a quote from the first of these titles, in which, after the exhumation of Mr. Appleton's body, someone says "arsenic, wasn't it?" and receives the reply, "No, strychnine, I think. Not that it makes any difference."

How Does Your Garden Grow is more interesting, for it makes oblique reference to two further real-life strychnine poisoning cases. Christie had clearly given thought to the ideal way of administering such a horribly bitter substance. She first brings out the possibility that a cachet was used. This was an obsolete kind of medication consisting of an unpalatable drug wrapped in a folder of rice-paper, and is closely related to the gelatine capsules used by the Waterloo murderer Thomas Cream in the 1890s. She then reveals the true answer, which is—inside an oyster. This ingenious solution does not have any parallel in the world of real crime, but there is a faint echo of another case in the way in which the culprit tries to incriminate the Russian servant-girl; she hides a packet of strychnine under the mattress in her room. This harks back to the trial of Walter Horsford in 1898.[8] Horsford poisoned his cousin, Annie Holmes, a widow with whom he was having an affair. The women laying out the body found two packets of strychnine under the mattress, together with a letter from Horsford making it clear that he had sent it to Annie Holmes described as a harmless abortifacient, although the postmortem showed that she was not pregnant. A feature of the case was that the police had already searched the house no fewer than four times, and failed to find the strychnine.

This profile of the alkaloid in Christie's stories is not bad for a substance which by the time she was writing was out of use as a domestic poison; indeed, in the period during which she wrote, its appearance in her books was running almost neck-and-neck with its illicit usage in the whole of the real world. But like strychnine itself, fictional strychnine ends on a dying fall; whodunits by others reinforce the picture of strychnine reduced to a cypher, a handy shorthand for a poison that everyone has heard of, but no one can be bothered to know much about. *Stolen Strychnine*, by Belton Cobb (1949), is one of the better reads; third-rate rather than tenth-rate, but the plot is nevertheless tedious and dated.

The whole genre had been effectively lampooned in P.G. Wodehouse's short story *Strychnine in the Soup* from a few years before. Cyril Mulliner and his intended future mother-in-law are both so desperate to get hold of a copy of the eponymous book to find out how it ends ("Is there a Faceless Fiend? —There are *two* Faceless fiends!"—Thomas Cream and Jack the Ripper perhaps?) that they ransack the house to get hold of a copy and end up reading it together. The resulting accord is such that she changes her mind about him and gives permission for him to marry her daughter. Wodehouse had very possibly read *The Big Four* by Christie, a prime case of desperate plotting that did not in fact contain two faceless fiends, but might just as well have done. In no particular order, a man is poisoned by a fake doctor with what is at first thought to be strychnine but is found to be another alkaloid, Gelsemine, while a Chinese servant puts opium in his dinner and shoves him headfirst into the fire. It is not known whether Christie wrote this as self-parody.

While we are on the subject of Agatha Christie, perhaps it is worth describing a real-life poisoning case from her era. With its poison-pen letter, a stock-in-trade for Christie, it could almost have been penned by her, but the dénouement is sadder and more convincing than anything she was capable of writing.

No doubt Christie read with great interest the events at Horncastle, a small Lincolnshire coastal town, in 1934. The police investigation began with this letter:

To Inspector Dodson, Horncastle Police

Sir,

Have you ever heard of a wife poisoning her husband? Look further into the death (by heart failure) of Mr. Major, of Kirkby-on-Bain (Lincolnshire). Why did he complain of his food tasting nasty and throw it to a neighbour's dog, which has since died? Ask the undertaker if he looked natural after death? Why did he stiffen so quickly? Why was he so jerky when dying? I myself have heard her threaten to poison him years ago. In the name of the law, I beg you to analyse the contents of the stomach.

"Fairplay"

The author of this letter was never traced. But its accusations were sufficient to galvanise Dodson. He interviewed the doctor who had treated Arthur Major. This man had never seen Major when he was alive, but Mrs. Major had reported him subject to occasional fits. She had called the doctor to the house one night and told him that her husband had suffered another and died.

Dodson went to the council house in Kirkby-on-Bain where the stony-faced Ethel Major had lived with her lorry-driver husband. He had been a heavy drinker, and the home life of the Majors had been tempestuous, with an extraordinary amount of backbiting in the village. Ethel had instructed a solicitor to send letters to the neighbours. When Arthur found out about it, he went around to the solicitors' office and told them, "I'll wring her ------- neck."[9]

Dodson stopped the funeral and interviewed the next-door neighbour whose dog had died; together they dug up the small corpse. As he was leaving the Majors' house, Ethel Major said, "I'm not suspicioned, am I? I haven't done anything wrong."

The organs of both the dog and the human victim showed strychnine. But apart from her slip of the tongue at the threshold, there was no evidence linking Ethel Major to the poisoning, although another piece of circumstantial evidence was that her elderly father was a retired gamekeeper, renowned in the district for his skillful use of strychnine in game traps. When Scotland Yard were called in, her ready tongue further incriminated her. She told chief inspector Young that she blamed her husband's death on his addiction to tinned corned beef, which he always prepared himself. Young asked her if she was familiar with her father's reputation as a gamekeeper. She replied, "I didn't know where he kept his poison. I never at any time had any poison in the house, and I didn't know that my husband had died of strychnine poisoning." Asked why she had brought up the name strychnine, she said, "Oh, I'm sorry, I must have made a mistake." She then offered him a cup of tea. "You needn't be afraid of me," she told him, "I won't put anything in it." Earlier, he had found her wardrobe full of expensive dresses, and 20 pairs of unworn shoes. Despite her protestations, he established that she had sent her son out to buy corned beef.

Young then visited her father and asked him where he kept his strychnine. The 70-year-old man showed him a bottle of it in his bedroom, in a locked box for which there was only one key, the one he kept attached to his belt, and which was never out of his sight. This seemed a major difficulty. Young had Ethel Major arrested and searched her house. In a suitcase under the bed, he found another key identical with the one he had seen only a few hours before. The father sealed his daughter's fate by telling him that yes, he had forgotten to mention that he once had another key, but it had gone missing several years before.

The forensic evidence showed that the dead man had taken two separate doses of strychnine, about 48 hours apart. As the judge pointed out, this was highly puzzling; it was almost impossible to conceive of someone swallowing a second dose aided only by something as bland as corned beef. But this was the only weak point in the prosecution case, and Ethel was found guilty, although the jury did add a recommendation to mercy. Their grounds were not stated, but it was probably because Arthur Major had probably been conducting an affair with one of the neighbours.[10] But judges still followed the pathway laid down in Henry VIII's time of regarding poisoning as so despicable, because premeditated, that appeals were almost never allowed except on grounds of insanity. Ethel Major was hanged at Hull prison in December 1934,[11] and the forensic career of strychnine in the British Isles came to an end, more or less.

Endnotes

1. I am not sure if such a demonstration of the colour test was ever in fact given in an American court. Lloyd may have been influenced by the trial of Mary Freet for poisoning her husband in Ohio in 1863. The prosecution asked that the test be shown to the jury, but the judge ruled that "Scientific experiments would not enlighten the jury, since they might not have the knowledge to understand them and hence they might be misled thereby." Freet was acquitted (*Ohio Medical and Surgical Journal*, 1864, vol. XVI, p. 107).
2. Flannery, 1998.
3. Gerald, 1993.
4. Smith and Cook, 1934, which contains an extensive description of strychnine poisoning, does not mention it, nor does the *Beilstein Handbook of Organic Chemistry*, 2nd Supplement in a bibliography on strychnine and its salts with over 1000 citations, although it is mentioned in one or two specialist pharmacy sources. According to a biography of Christie (Morgan, 1984), Christie was obsessed with cataloguing, and spent much of her time drawing up lists of drug incompatibilities.

5. *Pharmaceutical Journal*, July 21, 1923, p.61.
6. Haining, 1995; Hart, 1990.
7. Gerald, 1993.
8. Thompson, 1925:325–328.
9. *Times*, 1–2 November 1934.
10. As well he might. In about 1915, Ethel Brown, as she then was, had had an illegitimate child who was brought up by her family. She concealed this fact from Arthur when they married, and it was not until several years later that someone in the village told him that the girl he thought was Ethel's younger sister was in fact her illegitimate daughter (Butler, 1973:92).
11. Gribble, 1957.

CHAPTER 18

Another Round of Pay Phone Hysteria

Chemistry! Would to God I could blot out the word!

John Uri Lloyd, *Stringtown on the Pike*, 1900

Strychnine has claimed its permanent place in human mythology. The very name triggers in most peoples' minds a morbid, or gleeful, frisson with overtones of Halloween, despite the fact that most know rather little about it. It is usually confused with arsenic, with cyanide, with nerve gas...the name is taken in vain.

As if by 1943 it were not already sufficiently notorious, strychnine was said to have been one of the drugs that contributed to the deterioration of Hitler's health after the Stalingrad defeat the previous year. Despite the fact that it was only one of dozens of drugs known to have been ingested by him towards the end of his life, the word "strychnine" was used by David Irving[1] as the heading to the chapter in his book describing how the fortunes of the Third Reich began to collapse. There is considerable doubt about the facts, however, as there is about many of the events of the Führer's last period.

Hitler was surrounded by competing sets of doctors administering their own remedies. An SS doctor, Dr. Grawitz, had since 1936 been prescribing "Dr. Koester's Antigas Tablets," which the Führer took to the extent of eight or sixteen a day. In 1944 a visiting doctor examined a tin and was appalled to find that they contained belladonna and "Extr. Nux. vomic. 0.04" (four percent, presumably). The visitor announced, "...Hitler had from time to time been imbibing nearly lethal doses....So dramatically could strychnine amplify the reactions of the nervous system that lockjaw could result from a normally harmless stimulus."[2] The size of the pills and the strength of the nux vomica extract are not recorded, but it seems unlikely that ingesting pills containing such a dose of nux vomica would in itself have caused poisoning unless the patient was both extremely constipated and taking the pills compulsively at the rate of far more than sixteen a day[3]; both scenarios being of course exceedingly likely in Hitler's case. He suffered from constipation and flatulence as a result of his vegetarian diet, and by taking a large number of pills he would have been placing himself in the position of one of the unfortunate Victorian patients poisoned by an accumulation of nux vomica powder in a static intestinal canal. Another account, however, says that the visiting doctor (named as Dr. Giesing) had misread the label on the antigas pills.[4]

The end of the war did not result in a diminution of strychnine's reputation. Despite the rarity of its use in the real world, it had become "meta-strychnine," the abstract concept more deadly than the real alkaloid, and which can be called upon at will by composers of conspiracy websites, saving them the trouble of having to think. Not all manifestations are sinister; some are entertaining. There is a rock band called "Strychnine," and in Amsterdam a tranvestite cabaret artiste called "Molly Strychnine," who clearly feels that the name chimes with a repertoire of louche material leaning heavily on songs by David Bowie and Berthold Brecht, which she performs with her musical accompanist, Fuckoffsky (Figure 21).

Subject: FW: Please be careful using payphones.

Hello, my name is Tina Strongman and I work at a police station as a phone operator for 911.

Lately, we've received many phone calls pertaining to a new sort of problem that has arisen in the inner cities, and is now working its way to smaller towns. It seems that a new form of gang initiation is to go find as many pay phones as possible and put a mixture of LSD and strychnine onto the buttons. This mixture is deadly to the human touch, and apparently this has killed some people on the east coast. Strychnine is a chemical used in rat poison and is easily separated from the rest of the chemicals. When mixed with LSD, it creates a substance that is easily absorbed into the human flesh, and highly fatal.

Please be careful if you are using a pay phone anywhere. You may want to wipe it off, or just not use one at all. Please be very careful. Let your friends and family know about this potential hazard. Thank you.

This piece of advice also circulated over the name of Sergeant Terence D. Murchison of the U.S. Air Force, and was distributed worldwide by e-mail during 1998–1999. It is merely one of a long series of urban myth communications featuring strychnine. Readers will by now know that strychnine is not poisonous enough by a factor of perhaps 1,000 to kill anyone by skin absorption.[5]

The owner of the address (tinastrongman@hotmail.com) maintained that he or she was not responsible for the message. To add to the confusion, a website devoted to uncovering fraudulent information[6] incorrectly stated that this return e-mail address had been deliberately added as an experiment to see how rapidly bogus information could spread on the Web. The whole episode is completely fatuous.

Other websites, such as The Vaults of Erowid ("plants & drugs, mind & spirit, freedom & law, arts & sciences"), have researched rumours that LSD has sometimes been cut with, or substituted by, strychnine. This would be more dangerous, if true, and is a long-standing urban myth. In 1971 S.H. Schnoll and W.H. Vogel[7] found that traces of a substance that behaved like strychnine had shown up in a thin-layer chromatography analysis of LSD obtained in Philadelphia. Thin-layer chromatography is not a suitable technique for positively identifying a substance, and they did not carry out the necessary further analysis. Some samples of heroin seized on the streets of Münster, West Germany, during 1972–1976 definitely contained up to 1.6% strychnine; not enough to kill, and presumably added as an adulterant so as to increase the apparent strength of the "hit."[8] But in his book

LSD: My Problem Child, Albert Hofmann, the Swiss chemist who first synthesised LSD, recalls being approached by the Basel police department one day in 1970 to identify a drug sample from which someone had died. According to him, the powder contained no LSD at all, but was pure strychnine.[9]

Despite the horrific symptoms, suicide by means of nux vomica or strychnine is not unknown. In France, the famous forensic scientist Orfila and a colleague investigated an early case on April 23, 1825.[10] A few years later, Mr. Ollier, surgeon to the Western Dispensary in England was called to attend an 80-year-old woman who had deliberately taken half an ounce of nux vomica powder. Her skin was covered in perspiration, her pulse was quick and faint, and she was tormented by thirst; but when given water to drink, she did not seem inclined to take any. Ollier gave her repeated doses of ipecac to induce vomiting. Before he could give her the first dose, she suffered a transient convulsion, and felt great trepidation, refusing to let go of her husband. Then the fits came regularly, each one lasting a minute or two, not dissimilar to epileptic convulsions and separated by intervals of remission during which she said that she was not in pain. The fourth of these made her so stiff that with all his strength, the doctor was unable to bend the pelvis on the thighs to get her back into her chair. From this she never recovered, and died of asphyxia.[11] Probably the most celebrated strychnine suicide was Arizona congressman John H. Fitzgerald, a former "forty-niner" gold rush miner, who took a fatal dose at Mill City (now Phoenix), on July 22, 1871. These days, suicide by strychnine is rare but not unknown. In 2001, South Warwickshire coroner's court was told how 79-year-old retired farmer John Harrison of Sutton-under-Brailes, who had failed to get over the death of his wife eighteen months previously, neatly folded his clothes and went to bed at the cottage where he lived after taking it.[12] To die alone of strychnine poisoning, surely among the worst ends that can await anyone.

Deliberate ingestion of strychnine in near-suicidal quantities is practised by members of the Free Pentecostal Holiness Church. This is a religious sect with 2,000 congregants located in the mountains of Tennessee, Kentucky and parts of Virginia and North Carolina, and was founded in the early 1920s by George Hemsley, who broke away from the Church of God based in Cleveland,

Tennessee when it stopped endorsing serpent-handling. Members of the church drive themselves into a state of mania by loud and chaotic singing and praying, and then test their faith by burning themselves, handling poisonous snakes, or less frequently by drinking strychnine. There have been 77 deaths from snakebites since the 1920s. One casualty was Spencer Evans, who in 1996 nearly died after being bitten by a rattlesnake and spent his 24th birthday in the hospital. "I done took 'em up. I still believe it's right. The Bible didn't say they wouldn't bite," he was reported as saying. Although there are no documented fatalities from members of the church taking strychnine, it is claimed, though not properly authenticated, that one church member took half a grain (34 milligrams) during a trance without any ill-effect: "In my stomach, it feels like cold water, you can sprinkle it on my tongue and it tastes better than honey."[13] Do not try this at home.

The Talbot Arms in Rugeley is now a theme pub, and no trace of its original plan remains, nor that of William Palmer's residence opposite, now a clothing warehouse with fluorescent striplights. Only his mother's grand house, converted to offices, remains opposite the churchyard where John Parsons Cook's vault, erected by public subscription, still faces out the ghost of William Palmer after a century and a half.

Strychnine poisoning is, however, a dormant volcano, not an extinct one. In March 2000, two people nearly died in Brisbane as a result of strychnine-laced paracetamol tablets. This was an extortion plot by Dennis Fountain, aged 62, who contaminated Panadol and Herron brand tablets on the shelves of Australian supermarkets and tried to extract money from the manufacturers, SmithKline Beecham and Herron Pharmaceuticals. Like Christiana Edmunds many years before, he tried to cover his trail by also poisoning himself and his wife. Ironically, the two members of the public who nearly died were a Brisbane doctor and his son. Fountain was detected and hanged himself in his cell before he could be brought to trial.

Modern analytical chemistry, using techniques such as chromatography and mass spectrometry, is so sensitive that there is no chance of deliberate strychnine poisoning going undetected these days once an analysis is undertaken. But paradoxically, as poisoning becomes a rarer crime, there is an increasing risk that it may go undetected because every possible explanation except deliberate homicide occurs to the medical practitioner.

Such was the case in 1995, when an Indo-Canadian woman, Parvesh Dillon, fell violently ill and died four days later in a Toronto hospital. The cause of her mysterious death went unexplained until eighteen months later when an alert insurance investigator realised that another claim was being made following the death of her husband's business partner. As a result, Sukhwinder Singh Dillon was belatedly tried for two murders. Witnesses were called from India, and it appears that he might have obtained his strychnine supplies from the subcontinent.

Accidental poisoning from strychnine-containing herbal remedies is not infrequent, and Chinese herbal remedies can be dangerous. Nux vomica seeds are called Maqianzi, and a case was reported from Hong Kong in 2002 of a 42-year-old woman partially paralysed for several hours after an overenthusiastic prescription of it by a herbalist.[14]

Strychnine's position as the bitterest known substance faded with time. In 1955, Friedhelm Korte of the University of Hamburg isolated from plants of the gentian family a substance that he called Amarogentin. He found that its bitterness could be detected at a dilution of one in 58 million, which makes it about a thousand times more bitter than the more famous alkaloid.[15]

Strychnine is no longer a hot scientific research topic. The number of citations to it in *Chemical Abstracts* has been running at a modest 50 or so a year for at least the past 30 years. Most of these researches deal with its use as a tool for studying nerve impulses, or are improved analytical methods for detecting it, alongside many other drugs, in forensic samples. In the early 1990s, though four different groups of chemists disclosed new syntheses based on forty years of development of new synthetic methods since Woodward's time. One of these achieved a yield of nearly 10%, an improvement over Woodward by a factor of 100,000, while another group reduced the number of stages from Woodward's twenty-eight to fifteen.[16]

Successive legal restrictions mean that strychnine is now very difficult to get hold of. Throughout the 1960s there were complaints in the columns of British newspapers about the severity of newly introduced restrictions. Gamekeepers and others were getting supplies clandestinely from the doctors, sometimes via the local police. In 1969, the Council of the Pharmaceutical Society reminded its members that sales were allowed to doctors and vets only for their treatment of patients or animals, and that it was illegal to sell it even to these professionals for the purposes of killing vermin. Then, as strychnine effectively vanished as a human drug, even this loophole closed up. It is now prohibited for use against birds, rats and mice, and its sole permitted use is for poisoning burrowing mammals; moles in the UK and various kinds of gophers in North America. For this purpose, strychnine-impregnated earthworms or other bait is introduced into the runs, either by hand, or in Canada by using an implement called the "Gopher Getter Jr." Staff using it receive special training. In the UK it may only be used on commercial agricultural or horticultural land where public access is restricted, and not in public parks.[17] For some reason it can, however, be used on golf courses. A recent report says that moles are getting out of control in the UK, partly because of a shortage of strychnine from India where the nux vomica trees have been over-harvested.[18]

In parts of the world where a less restrictive regime operates, strychnine pesticides still pose danger to pets and wildlife. In general, it seems that the further one gets from Rugeley, the easier it is to get hold of. Until 1998 at least, anti-gopher products containing 0.5% strychnine were still on open sale in Washington State, and were causing the deaths of numerous domestic pets. This led to proposed changes in the legislation and increased restrictions. The Canadian government banned concentrated strychnine in 1992, but is currently under pressure from farmers to reintroduce it because of the gopher menace.

One of the first people to seriously consider the question of how plants make natural products like strychnine, was Robert Robinson in a 1917 paper. From the first, he recognised that the general principles by which the plants operated would have to obey the rules of organic chemistry as worked out in the laboratory, despite the fact that plants operate using the highly complex enzymes rather than laboratory reagents. Isotopic labelling experiments since the second world war have made it possible to trace each fragment of the molecule as the plant assembles it, and we now know a lot about the process. A question that rather surprisingly remains essentially unanswered at the present time is, what function do strychnine and other alkaloids perform in the plant? Not all plants produce alkaloids, but those that do expend a lot of their metabolic energy making these extraordinarily complex substances, some of which are capable of killing, some of giving hallucinations, some of sending to sleep, and some having no discernible biological properties. Why do they do it?

There have been two main theories. One maintains that alkaloids give the plants a definite ecological advantage by poisoning predators. The other theory is that alkaloids are essentially waste products using up excess stocks of "unwanted" nitrogen metabolites.

The philosophical grounding for this question, expressed as such a dichotomy, is rather shaky. In the Darwinian universe, plants to not have to have a purpose, stated in this anthropocentric way, for anything that they do. The question could be rephrased as follows; given that some plants have evolved the enzymes and metabolic pathways to produce these complex nitrogen-containing metabolites known as alkaloids, does their possession give the plants any evolutionary advantage? It seems clear that some alkaloids at least fulfill a useful purpose; on the other hand, some of them are useless.

In the constantly changing chemical battleground that is biological evolution, plants and their predators are constantly seeking out and probing each others' defences. Some plants have adapted to the challenge of being eaten by regenerating their foliage very quickly; others by developing hard waxes on the leaves, others by producing chemicals like alkaloids that may interfere with the metabolism of their attackers. It is not surprising that at any instant of geological time, such as the present, some of these scatter-gun mechanisms are no longer useful, or never were. But if the sum total of the defences that a plant species puts up become insufficient because of changed circumstances, it will gradually disappear, or adapt.

So the plants, only some of which produce alkaloids, are part of an extraordinarily complex matrix of organisms each striving for survival, each trading off the others. The millipede *Polyzonium rosalbum* feeds on grasses containing the alkaloid loline, absorbs it into its own body and converts it into a highly toxic metabolite, which then becomes part of its own defence secretions, making it poisonous for other animals to eat. Then there is the drugstore beetle. The insignificant dun-coloured *Stegobium paniceum*, an eighth of an inch long, lives in cupboards and pantries, where it passes its time quietly. For those who wish to eradicate an infestation, the recommended technique is to use a vacuum-cleaner. Pesticides are not a lot of use. It is said to eat "almost anything except cast-iron" and has been reported to thrive on opium, cayenne pepper or strychnine.[19]

Of the 200 or so *Strychnos* species recognised by modern botanists, about half have been investigated chemically; strychnine and brucine are the commonest alkaloids, but dozens of others have now been identified in various plants of the genus since those early days. In all, more than 20,000 alkaloids are now known. Not all come from plants; most spider and snake venoms contain nitrogen and can be considered alkaloids, and some of the most powerful recent isolates have been toxins produced by marine animals such as sponges.

It is estimated that one-tenth of the plant species in the world are used medicinally. When they are investigated carefully, it is found that of these about one-quarter show definite activity, due to known or to new drugs; the other three-quarters seem to have undeserved reputations, or have perhaps been mistaken for something else. Eighty percent of the world's population relies entirely or almost entirely on traditional medicine. Even in the West, about one-third of new medicines are derived from natural sources, with or without help from the synthetic chemist. Out there in the Amazonian rainforest, or deep in the ocean, there are certain to be hitherto undiscovered drugs with remarkable properties.

But to assume that everything natural is inherently good is a cardinal error. Strychnine never fulfilled its promise as a nineteenth-century wonder drug, and caused far more misery than could ever have been justified by its medicinal usefulness. Quite why the medical profession continued to believe in its benefits long after all the evidence was against, is something that this book has been unable to satisfactorily explain. But its availability did provide a great deal of more or less innocent amusement to the newspaper-reading public of two centuries.

Endnotes

1. Irving, 1990, Ch.28; Maser, 1971; 329–330.
2. Irving, 1990.
3. At Hitler's request, "Several gross" were supplied (Irving, 1990); if this refers to several gross tins, he could have been taking the pills in vast numbers. If, for the sake of illustration, they were 5-grain (300-mg) pills containing 4% of nux vomica extract, itself containing 7% strychnine, a fatal dose would have been contained in about 50 pills. Maser says that he was supposed to take 2–4 a day, but frequently exceeded this.
4. Eberle and Uhl, 2005; 164 (footnote).
5. The belief that some poisons are powerful enough to kill at the slightest touch goes back well before the unfortunate Edward Squyer in Chapter 7, although no one seems to believe nowadays in poisons that can kill by sight alone, as mentioned in the prologue. At an exhibition of poisonous fungi in France in the 1990s I had my wrist slapped (literally) by the curator for picking up an Amanita toadstool. It took the twentieth century to make the nightmare come true by inventing nerve gases, liquids that can indeed kill by skin absorption of a small droplet.
6. www.crimes-of-persuasion.com. See also www.urbanmyths.com.
7. New England Journal of Medicine, 1971, Vol. 234:791.
8. Bohn, Schulte and Audick, 1977.
9. Hofmann, 1979:8.
10. Lancet, 1825:56.
11. Lancet, 1837:855.
12. Cotswold Journal, 30 August 2001 He appeared to have kept the strychnine for many years (letter to the author from Michel F. Coker, Coroner).
13. McNeil, 1989; Augusta Chronicle online, June 28, 1996.
14. Chan, 2002.
15. Chemische Berichte, 1955, 88:704.
16. Beifuss, 1994.
17. Shenker, 1992. At the time of going to press, all strychnine use whatever had just been banned throughout Europe.
18. Guardian, March 4, 2003:11.
19. The beetle's ability to eat strychnine is one of those frequently-cited facts for which the original observation is difficult to track down. See for example Gaul, 1953.

Bibliography

Accum, Frederick (1820), *A Treatise on the Adulteration of Food, etc.*, London; Longman, Hurst, Rees, Orme and Brown.

Acton, William (1870), *Prostitution, Considered in Its Moral, Social and Sanitary Aspects in London and Other Large Cities*, 2nd ed., London; J. Churchill & Sons.

Albert, Adrien (1987), *Xenobiosis; Food, Drugs and Poisons in the Human Body*, London; Chapman & Hall.

Altick, Richard D. (1970), *Victorian Studies in Scarlet*, New York; W.W. Norton and Company.

Anon. (1823a), *Affaire Castaing. Accusation D'Empoisonnement, Publié par un Temoin.* Paris; Le Stenographe Parisien, chez Delongchamps, Delaunay, Ponthieu, Martinet et Sanson.

Anon. (1823b), *Procès Complet d'Edme-Samuel Castaing, Docteur en Médécine*, Paris; Chez Pillet Ainé.

Anon. (1823c), *An essay on Mineral, Animal and Vegetable Poisons, in which the Symptoms, Mode of Treatment and Tests with the General Morbid Appearances Are Concisely Detailed, to which Is Added the Means to Be Employed in Cases of Suspended Animation*, 3rd ed., Southwark; Cox.

Anon. (1856), *Illustrated Life and Career of William Palmer of Rugeley*, London; Ward Lock.

Bardsley, James (1830), *Hospital Facts and Observations, Illustrative of the Efficacy of the New Remedies Strychnia, Brucia, Acetate of Morphia, Veratria, Iodine etc.*, London; Burgess & Hill.

Barrett, C.R.B. (1905), *The History of the Society of Apothecaries of London*, London; Elliott Stock.

Beaconsfield, Lord (Benjamin Disraeli) (1852), *Lord George Bentinck; a Political Biograph*, London; Colburn.

Beifuss, Uwe (1994), *Angewandte Chemie International Edition*, 33:1144–1149.

Bell, Jacob (1841), Historical sketch of the progress of pharmacy in Great Britain, *Pharmaceutical Journal*, 1841, vol. 1.

Bennett, Angelo (1856), *Verbatim report of the Trial of William Palmer, transcribed from the shorthand notes of Mr. Angelo Bennett*, London; J. Allen.

Berger, Natalia (Ed.) (1995–1997), *Jews in Medicine*, Tel Aviv and Philadelphia; Beth Hatefutsoth/Jewish Publication Society.

Bernays, Albert J. (1855), *First Lines in Chemistry; A Manual for Students*, London; Parker and Son.

Berthelot, M. (1872), *Traité Elémentaire de Chimie Organique*, Paris; Dunod.

Birkenhead, Earl of (1935), *Frederick Edwin, Earl of Birkenhead*, 2 vols., London; Thornton Butterworth.

Blundell, R.H. and Seaton, R.E. (1929), *Trial of Jean Pierre Vaquier, Notable British Trials Series*, London; William Hodge and Company.

Bohn, S.G., Schulte, E., and Audick, W. (1977), *Archiv. der Kriminologie*, 160, 27–33.

Boo-Chang Cai and coworkers, (1990), *Chem. Pharm. Bull. (Japan)*, vol. 38, p. 1295.

Bridges, Yseult (1956), *How Charles Bravo Died*, London; Jarrolds.

Brock, William H. (1997), *Justus Von Liebig, The Chemical Gatekeeper*, Cambridge; Cambridge University Press.

Browne, G. Lathom and Stewart, C.G. (1883), *Reports of Trials for Murder by Poisoning*, London; Stevens and Sons.

Bouchardon, Pierre (1925), *Le Crime du Château de Bitremont*, Paris; Albin Michel.

Buckingham, John (2004), *Chasing the Molecule*, Stroud, Glos.; Sutton.

Busquet, Paul (1927–1936), *Les Biographies Medicales*, 5 vols., Paris; J.-B. Ballière et Fils.

Burney, Ian A. (2000), *Bodies of Evidence; Medicine and the Politics of the English Inquest, 1830–1926*, London; Johns Hopkins University Press.

Butler, Ivan (1973), *Murderers' England*, London; Robert Hale.

Campbell, R. (1747), *The London Tradesman, Being a Compendious View of All the Trades, Professions, Arts, both Liberal and Mechanic, Now Practiced in the Cities of London and Westminster*, London; T. Gardner, facsimile edition by David and Charles, 1969.

Campbell, John (1881), *Life of John, Lord Campbell, Edited by his Daughter*, London; John Murray, 1881, 2 vols.

Chan, T.Y.K. (2002), *Human and Experimental Toxicology*, 21, 467–468.

Chapman, G.T.L. and others (1995), *A New Herball by William Turner*, Cambridge; Cambridge University Press.

Chevalier, François Jacques (1820), *Dissertation sur l'usage et l'abus de l'émétique, dans quelques cas de medicine pratique*, Paris.

Chevallier, M.A. (1850), *Dictionnaire des Adultérations et Falsifications des Substances Alimentaires*, Paris; Béchet Jeune.

Chevallier, M.A. (1875), *Dictionnaire des Adultérations et Falsifications des Substances Alimentaires*, 4th ed., Paris; P. Asselin.

Chevers, Norman (1870), *A Manual of Medical Jurisprudence for India*, Calcutta; Thacker, Spink & Co.

Chopra, Sir Ram Nath and others (1949), *Poisonous Plants of India*, Calcutta; Indian Council of Agricultural Research, India Press.

Christison, Robert (1829), *A Treatise on Poisons*, 1st ed., Edinburgh; A. and C. Black.

Christison, Sir Robert (1885), *The Life of Sir Robert Christison, Bart, Edited by His Sons*. Edinburgh; W. Blackwood and Sons, 2 vols.

Clark, M. and Crawford, C. (1994), *Legal Medicine in History*, Cambridge; Cambridge University Press.

Clusius, Carolus (1567), *Aromatum, et Simplicium Aliquod Medicamentorum Apud indos Nascentium Historia (étant la traduction latine des Coloquios dos simples et drogues e cousas medicinais da India de Garcia da Orta)*, Facsimile ed. by M. de Jong and D.A. Wittkop Koning; Niewkoop, Netherlands; O.B. de Graaf, 1963.

Coatsworth, Dr. and others (1718), *Pharmacopoeia Pauperam, or the Hospital Dispensary*, London; T. Warner.

Coley, Noel G. (1991), *Alfred Swaine Taylor, MD, FRS, Forensic Toxicologist; Medical History*, 35, p. 409.

Colquhoun, Patrick (1800), *A Treatise on the Commerce and Police of the River Thames, etc.* London; Joseph Mawman.

Copeman, W.S.C. (1967), *The Worshipful Society of Apothecaries of London: A History 1617–1967*. Oxford, Pergamon Press.

Cranfield, G.A. (1978), *The Press and Society from Caxton to Northcliffe*, London; Longman.

Cripps, E.C. (1927), *Plough Court; The Story of a Notable Pharmacy, 1715–1927*, Allen and Hanbury.

Culpeper, Nicholas (1653), *The English Physician Enlarged*, London; Peter Cole.

Curling, Jonathan (1938), *Janus Weathercock; The Life of Thomas Griffiths Wainewright*, London; Thomas Nelson and Sons.

de Meyrignac, H.P. (1859), *De l'Empoisonnement par la Strychnine*, thesis, Paris.

Desportes, Eugène-Henri (1808), *De La Noix Vomique*, thesis, Paris.

Duke, James A. (1985), *Handbook of Medicinal Herbs*, Boca Raton; CRC Press.

Dumas, Alexandre (1907–1909), *My Memoirs*, Transl. by E.M. Waller, London; Methuen, 1907, 6 vols.

Dupré, Ernest and Charpentier, René (1909), *Les Empoissoneurs*, Lyon; A. Rey et Cie.

Eaton, Horace Ainsworth (1936), *Thomas De Quincey*, Oxford; Oxford University Press.

Eberle, Henrik and Uhl, Matthias (2005), *The Hitler Book*, Transl. by Giles MacDonogh, London; John Murray.

Fellows, James I. (1881), *A Few Remarks upon Fellows' Hypophosphites of Quinine, Strychnine, etc.*, London; Fellows.

Fenton, John Joseph (2002), *Toxicology; A Case-Oriented Approach*. Boca Raton; CRC Press.

Fielding, Steve (1994–1995), *The Hangman's Record*, 2 vols., Beckenham; Chancery House Press.

Flannery, Michael (1998), *John Uri Lloyd, The Great American Eclectic*, Carbondale; Southern Illinois University Press.

Fletcher, George (1925), *The Life and Career of Dr. William Palmer of Rugeley*, London; T. Fisher.

Flückiger, Friedrich A. and Hanbury, Daniel (1879), *Pharmacographia; A History of the Principal Drugs of Vegetable Origin Met with in Great Britain and British India*, London; Macmillan.

Fodéré, F.E. (1813), *Traité de Médicine Légale et d'Hygiène Publique, ou de Police de Santé Adapté aux Codes de l'Empire Français et aux Connaissances Actuelles*; Paris; Imprimerie de Mame.

Folsom, Charles (1909), *Studies of Criminal Responsibility and Limited Responsibility*, Privately printed, no place given.

Forbes, Thomas Rogers (1985), *Surgeons at the Bailey; English Forensic Medicine to 1878*, London; Yale University Press.

Fouquier de Maissemy (1802), *Avantages d'une Constitution Faible*, Aperçu medical, Paris; Imprimerie de Gillé fils.

Francis, John (1853), *Annals, Anecdotes and Legends, a Chronicle of Life Assurance*, London; Longman, Brown, Green and Longman.

Furber, Holden (1948), *John Company at Work; A Story of European Expansion in India in the Late Eighteenth Century*, Cambridge, Massachusetts; Harvard University Press.

Gaul, Albro T. (1953), *The Wonderful World of Insects*, London; Victor Gollancz.

Gerald, M.C. (1993), *The Poisonous Pen of Agatha Christie*, Austin, Texas; University of Texas Press.

Gerard, John, (1975), *Herbal, or General History of Plants, 3rd ed.*, modern edition, New York; Dover Publications.

Graham, T.G and Hofmann, A.W. (1852), *Report Upon the Alleged Adulteration of Pale Ales by Strychnine*, London.

Graves, Robert (1957), *They Hanged My Saintly Billy*, London; Cassell.

Gribble, Leonard (1957), *Famous Judges and Their Trials, a Century of Justice*, London; John Long.

Grigg, John (2002), *Lloyd George, War Leader 1916–1918*, London; Allen Lane.

Haining, Peter (1995), *Agatha Christie's Poirot, A Celebration of the Great Detective*, London, Boxtree/LWT.

Hamilton, James (2002), *Faraday; The Life*, London; HarperCollins Publishers.

Hart, Anne (1990), *Agatha Christie's Hercule Poirot*, London; Pavilion Books.

Hassall, A.H. (1876), *Food, Its Adulterations and the Methods for Their Detection*, London; Longmans, Green and Co.

Hazlitt, William Carew (1880), *Essays and Criticisms (by T.W. Wainewright), with some account of the Author*, London; Reeves and Turner.

Hervey, Charles (1892), *Some Records of Crime, Being the Diary of a Year, Official and Particular, of an Officer of the Thuggeee and Dacoitie Police*, London; Sampson Low, Marston and Company.

Hofmann, Albert (1979), *LSD: Mein Sorgenkind*, Stuttgart; Klett-Cotta Verlag.

Hogg, James (1895), *De Quincey and His Friends*, London: Sampson Low, Marston and Co.

Houlton, Joseph (1828), *Formulary for the Preparation and Employment of Several New Remedies, translated from the sixth edition of the Formulaire of M. Magendie*, London; T and G. Underwood.

Howes, Edwin, M.D. (1904), *A Treatise on Nux Vomica*, Drug Treatises No. VIII, Cincinatti, Ohio; Lloyd Bros.

Husemann, (1857), *Hufeland's Journal für praktischen Heilkunde*, 511.

Irving, David (1990), *Hitler's War*, New York; Avon Books.

Irving, Henry Brodribb (1918), *A Book of Remarkable Criminals*, London; Cassell.

Jalland, Pat (1996), *Death in the Victorian Family*, Oxford; Oxford University Press.

James, Frank A.J.L. (1991), *The Correspondence of Michael Faraday*, London; Institution of Electrical Engineers.

Kaye, John William (1843), *The Administration of the East India Company*, London; Richard Bentley.

Knelman, Judith (1998), *Twisting in the Wind; The Murderess and the English Press*, Toronto; University of Toronto Press.

Knott, George H. (1912), *Trial of William Palmer*, Notable English Trials Series; Edinburgh and London; William Hodge and Company.

Lawn, Brian (1963), *The Salernitan Questions*, Oxford; Oxford University Press.

Leeuwenberg, A.J.M. (1969), *The Loganiaceae of Africa VIII; Strychnos III*, Wageningen; H. Veenman & Zonen.

Lewis, Dave (2003), *The Rugeley Poisoner*, Stafford; Artloaf.

Lindop, Grevel (1981), *The Opium Eater; A Life of Thomas De Quincey*, London; J.M. Dent.

Mackay, Charles (1995), *Extraordinary Popular Delusions and the Madness of Crowds*, Modern ed., Ware; Wordsworth Reference.

Magendie, François (1813), *De l'Influence de l'Émétique sur l'Homme et les Animaux*, Paris.

Magendie, François (1822), *Formulaire Pour la Préparation et L'Emploi de Plusieurs Nouveaux Medicaments*, Méquignon-Marvis; Paris.

Marjoribanks, Edward (1929), *The Life of Sir Edward Marshall Hall*, London; Gollancz.

Mart, G.L. (1835), *Practical Observations on the Nature and Treatment of Nervous Diseases, with Remarks on the Efficacy of Strychnine in the More Obstinate Cases*, London; Churchill.

Maser, Werner (1971), *Adolf Hitler; Legende, Mythos, Wirklichkeit*, München u. Esslingen; Bechtle Verlag.

Masson, David (1914), *De Quincey*, London; Macmillan.

Maudsley, H. (1874), *Responsibility in Mental Disease*, London; Henry S. King.

McLaren, Angus (1993), *A Prescription for Murder; The Victorian Serial Killings of Dr. Thomas Neill Cream*, Chicago; University of Chicago Press.

McNeil, W.K. (1989), *Appalachian Images in Folk and Popular Culture*, Ann Arbor; University of Michigan Press.

Mitchell, John and others (1855), *Adulteration of Food, Drink and Drugs; Evidence Taken before the Parliamentary Committee of Enquiry*, London; David Bruce.

Mohr, James C. (1978), *Abortion in America, The Origins and Evolution of National Policy, 1800–1900*, New York; Oxford University Press.

Morgan, Janet (1984), *Agatha Christie; A Biography*, London; Collins.

Motion, Andrew (2000), *Wainewright the Poisoner*, London; Faber and Faber.

Myers, Robin, Harris, Michael and Mandelbrote, Giles (2001), *Under The Hammer; Book Auctions Since the Seventeenth Century*; London; Oak Knoll Press and the British Library.

Nadkarni, K.M. (1954), *Dr. K.M. Nadkarni's Indian Materia Medica*, 3rd ed., Bombay; Bombay Popular Prakashan Press, 2 vols., reprinted 1982.

Norman, Charles (1956), *The Genteel Murderer*, New York; Macmillan.

Olstead, J.M.D. (1944), *François Magendie*, New York; Schuman's Publishers.

Orfila, M.P. and Black, R.H. (1820), *Directions for the Treatment of Persons who Have Taken Poison*, 2nd ed., London; Longmans, Hurst, Rees, Orme and Brown.

Owen, D.J. (1927), *The Port of London, Yesterday and Today*, London; Port of London Authority.

Pariset, M. (1833), *Éloge de M. Vauquelin*, Mémoirs de l'Académie Royale de Médecine, vol. II, pp. 39–59. Paris; J.B. Baillière.

Pearce, John M.S. (2003), *Fragments of Neurological History*, London; Imperial College Press.

Pelletier, J. and Caventou, J.B. (1819). Sur un nouvel alcali végétal (la strychnine) trouvé dans la fève de Saint-Ignace, la noix vomique, etc., *Annales de chimie*, 10:142–177.

Pereira, Jonathan (1855), *Elements of Materia Medica*, 4th ed., London; Longmans, Brown, Green and Longmans.

Periquet, Charles (1854), *De La Strychnine*, thesis, Paris.

Pickering, Charles (1879), *Chronological History of Plants*, Boston; Little, Brown.

Pitt, R. (1707) *The Calamities of All the English in Sickness and the Sufferings of the Apothecaries from their Unbounded Increase, etc.*, London; John Morphew.

Porta, Giovanni (John) Baptista (1558), *Magiae Naturalis*, Naples, transl. anonymously into *Natural Magick* (London, 1658), modern facsimile ed. Basic Books, N.Y., 1957.

Rees, J. Aubrey (1923), *The Worshipful Company of Grocers: An Historial Retrospect*. London; Chapman & Dodd.

Remer, W.H.G. (1816), transl. Bouillon-Lagrange, E.J.B., *Police-judiciaire pharmaco-chimique*, etc., Paris; Caille et Ravier.

Requin, A.P. (1852), *Notice sur Fouquier*. Paris; Germer Ballière.

Ring, Carlyn (1980–1984), *For Bitters Only*, and update, 1984, Wellesley Hills, Mass.; Pi Press.

Robinson, Sir Robert (1976), *Memoirs of a Minor Prophet*, Amsterdam and London; Elsevier.

Rosenfeld, Louis (1999), *Four Centuries of Clinical Chemistry*, Amsterdam; Gordon and Breach.

Roughead, William (1938), *The Seamy Side*, London; Cassell and Company.

Roughead, William (1939), *Neck or Nothing*, London; Cassell and Company.

Russell, Colin A. (1983), *Science and Social Change 1700–1900*. London; Macmillan.

Schultes, Richard Evans (1995), *Ethnobotany, Evolution of a Discipline*, London; Chapman & Hall.

Secheyron and Daunic (1902), *Congrès Français de Médecine, 6th Session*. Toulouse; Edouard Privat.

Shenker, A.M., Ed., (1992), *Pest Control; A Reference Manual for Pest Control Staff*, London; ADAS/Local Government Management Board.

Shore, W. Teignmouth (1923), *Trial of Thomas Neill Cream, Notable British Trials Series*, London and Edinburgh; William Hodge and Company.

Shore, W. Teignmouth (1935) in *The Black Maria, or the Criminal's Omnibus*, Harry Hodge, Ed. London; Victor Gollancz/William Hodge and Co.

Shortt, John, M.D. (1867), *Trans Obstetrical Society of London*, vol. 9, p. 6.

Sigmond, George S. (1837), Lecture on Material Medica and Therapeutics, *Lancet*, 826.

Smith, John (1882), *A Dictionary of the Popular Names of the Plants Which Furnish the Natural and Acquired Wants of Man, etc.*, London; Macmillan.

Smith, Sydney and Cook, W.H.G., Eds., (1934), *Taylor's Principles and Practice of Medical Jurisprudence*, London; J.A. Churchill Ltd.

Smith, Roger (1981), *Trial by Medicine; Insanity and Responsibility in Victorian Trials*, Edinburgh; Edinburgh University Press.

Smith, J.C. and Hogan, B. (1988), *Criminal Law*, 6th ed., London:

Smyth, Frank (1980), *Cause of Death*, London; Orbis Publishing.

Steggall, John (1838), *Manual for Students Preparing for Examinations at Apothecaries Hall*, 9th ed., London; John Churchill.

Stuart, David (2004), *Dangerous Garden; The Quest for Plants to Change Our Lives*, London; Frances Lincoln.

Symons, Julian (1960), *A Reasonable Doubt; Some Criminal Cases Re-Examined*, London; The Cresset Press.

Talbot, W.A. (1911), *Forest Flora of the Bombay Presidency and Sind*, 2 vols. Poona; Government Photozincographic Dept.

Tardieu, Ambroise (1857), *Mémoire sur l'Empoissonnement par la Strychnine, Contenant la Relation Medico-légale Complète de l'Affaire Palmer*, Paris; J.B. Baillière.

Taylor, Alfred Swaine (1856), *On Poisoning with Strychnia; With Comments on the Medical Evidence Given at the Trial of W. Palmer for the Murder of J.P.Cook,* London.

Taylor, Alfred Swaine (1859), *Medical Jurisprudence,* 2nd ed., London; John Churchill.

Taylor, Alfred Swaine (1875), *Poisons, in Relation to Medical Jurisprudence and Medicine,* 3rd ed. London; J. & A. Churchill.

Thomas, J.M. and Thomas, W.J. (1997), *Principles and Practice of Heterogeneous Catalysis,* Weinheim; VCH.

Thompson, C.J.S. (1925), *Poison Mysteries in History, Romance and Crime,* 2nd ed., London; The Scientific Press.

Thornbury, Walter (1879), *Old Stories Re-told,* London; Chapman & Hall.

Todd, Lord and Cornforth, J.W. (1976), *Obituary Notices of Fellows of the Royal Society.*

Tomlinson, Thom (1994), *India Pale Ale, Part I; IPA and Empire, Brew Your Own Magazine,* vol. 2, No. 2.

Trousseau, A. and Pidoux, H. (1839), *Traité de Thérapeutique et Matière Médicale* [sic], Société Typographique Belge.

Trousseau, A. and Pidoux, H. (1875–1877), *Treatise of Therapeutics,* 9th ed., Transl. by D.F. Lincoln. London; Sampson Low, Marston, Searle and Rivington (publ. 1881).

Udwadia, F.E. (1996), *The Oxford Textbook of Medicine,* 3rd ed., Oxford; Oxford University Press.

Vandenbussche, L. and Braeckman, P. (1976), *Historische Vergiftingsdrama's in Europa,* Handzame, Netherlands; Familia et Patria.

Weir, Alison (1999), *Eleanor of Aquitaine,* London; Johnathan Cape.

Williams, Trevor I. (1990), *Robert Robinson, Bibliographical Memoirs of Chemist Extraordinary,* Oxford; Clarendon Press.

Williamson, W.H. (1930), *Annals of Crime; Some Extraordinary Women,* London; G. Routledge.

Wilson, R. Macnair (1926), *The Beloved Physician; Sir James Mackenzie,* London; John Murray.

Wilson, Colin (1989), *Written in Blood; a History of Forensic Detection,* London; Equation Books.

Woods, Oliver and Bishop, James (1983), *The Story of the Times,* London; Michael Joseph.

Zimmerman, David (1994), *The Encylopaedia of Advertising Tins; Smalls and Samples,* Pleasant Plain, Ohio; published by the author.

Index

A

Abercromby, Helen, 113–116, 118
Abercromby, Madeleine, 113–114,
    117–118
Abercromby, Mrs., 113–114
Abley, Mr., 162
Abortion, 128, 140, 192–194,
    197–198, 215, 254
Abscess, 24
Academie des Sciences, French, 48,
    51, 54
Academy of Medicine, French, 1
Acetone, 223
Acetylene, 222
Acetylcholine receptors, 13
Acidulated draught, Campbell's, 149
Aconitine, 93–95
Adamsky, Alice, 92
Adelard of Bath, 21
Adud al-Dawlah, Caliph, 20
Adudu Hospital, 20
Adulteration
    of Angostura, 59–62
    of beer, 211–215
    of foods, 187, 211
    of LSD, 261–262
    of medicinal strychnine, 62, 70, 73
    of nux vomica, 30, 70
Adventure of the Egyptian Tomb, The,
    253
Advertising Tins, Encyclopaedia of,
    216

Ahwaz, 20
Aiche Rhorab, 20
Aisne, 2
Aitchison, Dr., 207
Albion Inn, 131
Alcohol, 7, 11, 60, 72, 207
Alderson, Baron, 117, 142
Alexander VII, Pope, 84
Alfalfa, 211
Alfreton, 185
Alienists, 178–179, 191
Alimentary canal, 8, 25, 54, 72–74,
    77–78, 88, 105, 204–205, 260
Alkaloids, 50–52, 67, 266–267
Allen & Hanbury, 3, 53, 93, 203
Alliance Insurance Company, 114
All Products Unlimited, Inc., 208
Allsopps brewers, 211–212, 215
Aloes, 36, 204
Amarogentin, 264
Amaurosis, 75
Amenorrhoea, 69
Amsterdam, 260
Anaesthethists, use of strychnine by,
    79
Analysis, 221
Angina, 158
Anglesey, Marquis of, 126
Angostura, 58–60
    false, 58–60
Angosturine, 60
Animal magnetism, 2

Annales d'Hygiene Publique, 91
Antiaris, xix
Anticancer drugs, 74
Antidotarum, 20
Antigas pills, 259–260
Antimony potassium tartrate (tartar
    emetic)
    analysis for, 162
    detection in body, 147
    as emetic, 40, 104–105, 114
    interferes with detection of
        strychnine, 169
    poisoning with, 107, 118–119,
        146–147, 151, 160
    practical joke with, 134
Aphrodisiacs, 198, 207–208
Apoplexy, 89, 151, 153
Apothecaries
    Company/Society of, 38, 44, 89
    Dutch, 59
    examinations, 43, 45
    faithfull, 26
    German, 58, 72
    incompetent and overpaid, 41–42
    struggle for recognition, 37–39
Appleton, Mr., 253
Aqua Tofana, 92
Arabs; Arabia, 19–20, 24, 26, 248
Arachnitis, 106, 157
Arizona, 262
Armée, Grand, 10
Army, British, 29
Army, Republican, 2
Arnaud, E.R., 1
Aromatum Historia, 21
Arrack, 207
Arsenic
    habituating poison, 17
    medicinal use of, 76
    poisoning with, accidental, 92,
        111–112, 211
    poisoning with, deliberate, 89–91,
        95, 119, 183, 199
    regulation of, 92, 183
    thuggees, use by, 84–85
Arsenic Act, 1851, 92
Attention deficit disorder, 173
Atropine, 23
Austria, 26, 59

Autobiography, Robinson's, 228
Autopsy, *see* post-mortem
Avicenna, 20, 24–25, 40

B

Back, arching of, *see* Opisthotonos
Badgers, 211
Baghdad, 20
Bailey, Philip John, 207
Baird, John Logie, 249
Baker, Mr., 66
Ballet, Auguste, 99, 101–108, 248
Ballet, Hippolyte, 99–105
Bamford, Dr., 131–133, 138
Bank of England, 115
Banks, Sir Joseph, 59
Barbiturates, 54
Barnes, Lavinia, 132–133
Barrie, J.M., 167
Barrois, M., 248
Bart's, *see* Saint Bartholemew's
Basel, 262
Bass, brewers, 212, 215
Bate, George, 128, 139, 159
Batier, xvii
Batley's solution, 134
Battle of Britain, 239
Battle's vermin killer, 184
Baudin, Captain, 48
Baxter, Sidney, 187–189
Beard, Dr. and Mrs., 187–190
Beecham, 42
Beer, strychnine/brucine in, 211–215,
    248
Belgium, 90, 253
Belladonna, 92, 175, 204, 252, 260
Bell, Edward, 184
Belle River, 193
Bell, Jacob, 44
Bell, Sir Charles, 54
Bentinck, Lord George, 182
Benzodiazepines, 54
Bergen, Jane, 128
Berlin, 226
Bethnal Green, London, 37
Betting book, Cook's, 135, 137, 154
Bevan, Timothy, 35
Bezar, 23

Biarritz, 229
Big Four, The, 253, 255
Bilious pills, 186
Birkenhead, Lord, 242
Biological survey, US, 211
Birds, poisoning of, 25–26, 48, 265
Birmingham, 24, 167–168, 202
Bitremont, Chateau de, 90, 253
Bitrex, 219
Bitterness, 4, 101, 107–108
    of beer, 214
    of strychnine, 14, 52, 90, 231
Bitters, 204, 214
Bitter yellow substance, 49–51
Blackmail, 195, 215
Black cohosh, 216
Blackfriars, London, 41–42
Bladon, Mr., 162
Bladon, Prof. Peter, 235–236
Blanchard's Hotel, Quebec, 194
Bland, Mr., 230–231, 233
Blandois, Rigaud, 117
Bleak House, 87
Bleeding, 4, 102
Blistering, 4, 8, 9, 11
Blood, 4, 47–48, 74, 170, 171, 204
Blood–brain barrier, 12–13
Blue Anchor Hotel, 230–231, 233
Blue Geranium, The, 253
Boar, poisoning of, 147
Bocarmé, Count Hippolyte de, 90,
    253
Boils, 4
Bois du couleuvre, 22, 50, 60
Boleyn, Anne, 84
Bombay, 30, 84, 209
Bonplandia trifoliate, 58
Book, Robinson's, on strychnine, 228
Boone County, Illinois, 193
Booth, secret agent, 242–243
Borgias, 98
Boston Globe, The, 235
Boulogne, 115
Bourgeoisie, 182
Bowie, David, 260
Bowles, Henry and Mrs., 185–187,
    210–211, 218
Bowskill family, 185
Boyle, Robert, 40–41

Braconnot (Henri), 49–51
Bradford, poisoning of 1858, 92
Brain
    congestion of, 157
    Helen Abercromby's taken out, 118
    injury to, 4
    nerve impulses in, 5–6
    studies of, 47
Bravo, Charles, 109
Breadsall Church, 243
Brecht, Berthold, 260
Bretonneau, Dr., 68–69
Breweries, 211–215
Brewers' druggists, 214
Brewing Malt Liquors, 214
Bricheteau, Dr., 9
Bridges, Mr., 161
Brighton, 187–189
Brighton Times, 199
Brisbane, 263
Bristol, 156
Britain, *see also* England
    capitalism in, 76, 111
    commercialisation of strychnine in,
        33, 183
    forensic career of strychnine ends
        in, 257
    forensic science development in, 89
    medical opinion in, 76
    patients, dying, treatment with
        strychnine in, 80
    perfidious canting, 241
    social changes in, 63
    Victorian age in, 182
British Dyestuffs Corporation, 224
British Medical Journal, 168, 170,
    172–173, 205
British Pharmacopoeia, 204, 208
Britland, Mary-Ann and Thomas,
    184–185
Broadmoor Hospital, 191–192
Brodie, Sir Benjamin, 144–145, 153,
    165, 183
Bromo salts, 230–231, 233
Brookes (Palmer), Annie, 127, 131,
    142, 155, 161–162, 168–169
Brookes, Colonel, 127
Brooks, Mrs. Anne, 130
Brown, Ethel, 258

Brucea antidysenterica, 59–60
Brucea ferruginea, 59–60
Bruce, Jacques, 59
Brucine
   Back-street trade in, 217
   byproduct of strychnine
     manufacture, 62
   colour reactions of, 62
   Dumas and, 106, 248
   contamination of strychnine with,
     73, 75, 135, 195
   drug on the market, 213
   ignorance of, 214
   isolation of, 57–62
   metabolism of, 197
   naming of, 60–61
   occurrence, xix
   perfumes, use in, 217
   poisonous properties of, 60, 248
Buck and Rayner's drugstore, 193
Burdett, Ann, 111
Burgess, John, 127
Burial clubs, 88–89
Burion, M., 7–8
Burma, 29
Burns, Robert, 234
Burton ale scare, 211–214, 248
Burton-on-Trent, 211, 215
Bury, 90
Butterworth, John, 185
Button, Nathaniel, 163
Buzzards, poisoning of, 25
Byfleet, 230
Bynin liquid malt, 203
Byno hypophosphites, 203

C

Cabalistics, 41, 66
Cachets, 254
Calcutta, 30, 33, 66, 209
California, 211, 229
Calomel, 77, 131
Camberley, 186
Cambridge, England, 238
Cambridge, Massachusetts, 202
Caminarine, 62
Campbell, Lord (Sir John)
   beer adulteration and, 218
   coroners and, 88

   media and, 168
   Palmer trial and, 143, 149, 160, 162
   poisons legislation and, 182–183
   Stafford, opinion of, 164
   Wainewright case and, 115–116
Campbell, R., 41
Camphor, 35
Canada, 192–194, 264–265
Caniram, 29
Capitalism, Victorian Britain, 78
Carbazole, 222
Cardiac failure, strychnine for, 205
Caribbean Mystery, A, 252
Cartel system, expert witnesses, 142
Castaing, Dr. Edme-Samuel, 73,
   99–108, 123, 134, 172, 248
Catalysis, 172
Catatonia, 67
Cats, poisoning of, 26, 147, 176,
   188–189
Caulthrop, Benjamin, 188
Caventou, Joseph-Bienaime, 50–53,
   55, 59–62, 67, 92, 119, 164,
   221–222, 228
Cayenne pepper, 129, 267
Cerebellum, 12
Cerebral congestion, 118
Ceylon, 22
Chakravarti, R.N., 223
Chancre, 157
Charcoal, 54
Charite, Hôpital de la, 2, 7, 9–12
Château d'If, 247
Chelsea Physic Garden, 41
Chemical Abstracts, 264
Chemical messengers, 12
Chemical modification, of strychnine,
   53, 222, 225–226
Chemicals, foreign, 72
Chemical Society of London, 44
Chemical stability, of strychnine, 75,
   225–226
Chemical structure, of strychnine,
   227–228
Chemical tinderbox, 180
Chemists *see also* Apothecaries;
   Pharmacists, 40–42, 44, 53, 212,
   214
   German, 41, 222
   Lady Lirriper as, 249

Northern (USA), 250–251
prosecution and defence, changing
  places, 163
status, of 142
Chemistry
  crude speculations of, 172
  doctrines and laws of, 171
  foundation of, 1, 4
  instrumental techniques of, 234,
    237
  Liverpool College of, 42
  Manchester centre of, 223
  Oxford and, 225
  physical, of strychnine/potassium
    bromide, 253
  Robinson and, 222, 224–228
  Sherlock Holmes's knowledge of,
    249
  textbooks of, 213, 234
  Victorian ignorance of, 73, 163
  Woodward and, 234–237
Chemotaxonomy, 28, 223
Cheshire, Samuel, 139
Chesterfield, 222
Chesterton, G.K., 253
Chevalier, François Jacques, 105
Chevallier, M.A., 213
Chicago, 192, 201
Chicago Tribune, The, 193, 202
Chickens, poisoning of, 25, 48
Chicken, The, racehorse, 127, 130
Children
  diseases of, 5
  dosing of, 70
  Palmer's, 162
China, 19, 22
Chinchon, Countess of, 3
Chinese, abortions by, 215
Chloral, treatment for strychnine
    poisoning, 16–17, 54
Chlorine water, 53
Chloroform, 16
Chlorophyll, 50
Cholera, 67, 69, 89, 104
Cholesterol, 235
Christie, Agatha, 86, 247, 251–255
Christison, Sir Robert
  definition of poisons by, 71
  De Quincey, friendship with,
    122–123

indiscretion by, 95, 123
Palmer trial, evidence at, 145,
    147–148
public lectures by, 95
"poisons unknown to," 165
studies in Paris, 89, 122, 163
Treatise on Poisons by, 45, 71, 119
Chromatograhy, 251, 263
Chromium, 47
Church of God, 262
Cigue, 92
Cillière, Maurice, 10
Cinchona, 3, 27, 41, 57, 49–50, 57–58,
    203
Cincinnati, 251
Citric acid, 49
Ciudad Bolivar, 58
Civet cat, 30
Civil War, American, 250
Clarendon Laboratory, 224, 238
Clearing nut, 29, 51
Clear-up rate, 86
Cleveland, Tennessee, 263
de Clifford, Rosamund, 83
Clostridium tetani, 13, 144
Clover, Matilda, 194–195
Club money, 185
Clusius, Carolus, 22
Clutterbuck, Mr., 145
Cobalt nitrate, 231
Cobb, Belton, 254
Cobbett, William, 160
Cobras, 22
Cocaine, 204, 224
Cocculus indicus, 72, 214
Cochin China, 30
Cohosh, black, 216
Coleridge, Samuel Taylor, 120, 122
Colocynth, 77
Colour tests, 53, 91, 146, 148, 156,
    196–197, 221, 233, 237, 250–251
Colquhoun, Patrick, 35
Coming of Mr. Quin, The, 253
Complete Herbal and English
    Physician, 27
Concordia (ship), 35
Confectio sennae, 44
Confessions of an English Opium-
    Eater, 120
Conscientious objectors, 242

Considine, Dr., 15–17
Constantine the African, 20
Constipation, 8, 77, 205
Constitution, weak, 3
Consumption, *see* Tuberculosis
Continental Delivery Company, 212
Convulsions, 8–11, 71–72, 104, 107,
    143, 152–153, 155
Cook, John Parsons
    death of, 133–134
    explanations of death, 152–153,
        160, 172–173
    finances of, 129, 135–136, 150
    hypochondria of, 129
    illness of, 129–133
    inquest on, 146
    lifestyle, 151, 153
    penis of, 129, 155
    post-mortems on, 136–139, 169
    racing career, 128–129
    symptoms of, 107–108, 141, 144,
        146, 155–158, 179
    tonsils of, 129, 151
    vault of, 263
Copper salts, 111, 231
Cork, 161
Coroners
    incompetence of, 87–88, 139
    inquests, 111–112, 162, 139, 175
    juries, 65–66, 86
    professional position of, 87–88
Corporate blackmail, 215
Corrosive sublimate, 85
Cortisone, 235
Cotton, 35
Cottonroot, 216
Count of Monte Cristo, The, 247–248
Cream, Thomas, 93–94, 184, 192–
    198, 216, 229, 254–255
Creuzberg, Herr, 214
Cries of the Condemned, etc.
    (pamphlet), 168
Crimean War, 167
Criminal code, French, 103
Criminal Evidence Act, 141
Crisp, Inspector, 142
Crow fig, 29
Crows' bread, 20
Crows' eyes, 29
Crows, poisoning of, 26

Cruikshank, Isaac, 213
Crystallisation, crystal form, 51, 60,
    196, 226
Culpeper, Nicholas, 27
Cumulative poisons, 17, 67, 72–74
Cupping, 4
Curare, 13, 243–244
Curling, Thomas Blizzard, 143–144
Curry powder, 211
Curwen, Bennett, 83
Cusparia febrifuga, 58
Cutlers, 39
Cutler Street, London, warehouse, 36
Cyanide, 259

D

Dalby, Mr., 46
Dangerous drugs, warehousing of, 37
da Orta, Catarina, 23
da Orta, Fernando, 21
da Orta, Garcia, 21–23, 61, 246
Dark Ages, 19
Darlington, 184
Darndels, Mr., xvii
Darnel, 170
Darwin, Erasmus, xvii
Darwinian universe, 266
Datura, 23–24, 84–85
Davee, 31
Davy, Humphry, 172
Deafness, 66, 72
Death on the Nile, 253
Death penalty, abolition of, 176
DeBarenne, Dr., 12
Debauchery, 5
Degradation, of molecules, 225–226
Degeneration, 6
Deletion, strychnine from
    pharmacopoeia, plea for, 79
Delusions, insufficient for insanity,
    178, 190
Demons, 213
Denatonium benzoate, 219
Denglish, 212
Depressant, strychnine as, 203
Depression, 156–157
De Quincey, Thomas, 120–123, 181
Derby/Derbyshire, 185–186, 210,
    240–241, 243, 245

Desportes, Eugene, 49–51
Detroit, 215
Devi, Kujarani, 206
Devils, William Dove and, 173, 179
Devonshire, Mr., 137–138
Dhatoora, see *Datura*
Diabetes, 67
Dickens, Charles, 87, 115–117, 124
Digitalis, 175
Dillon, Parvesh, and Sukhwinder
    Singh, 264
Diseases, infectious, 4, 144
Dittany of Crete, 27
Dixon, Mary and Thomas, 184–185
Dobereiner, 180
Docks, London, 35–36
Doctors
    American, 78, 202, 204, 215, 217
    Arab, 20, 24
    British, 27, 37, 39, 67–69, 75–80,
        114–115, 118, 129–133, 137,
        143–145, 153–158, 175, 187,
        190–191, 203
    Eighteenth century, 4
    French, 2, 7, 18, 67–69, 76–78,
        99–106, 248
    German/central European, 7, 26,
        59, 259–260
    Illegible prescriptions of, 65
    Nineteenth century, 78
    South African, 13–17
    Strychnine new weapon for, xviii
Dodson, Inspector, 255–256
Dogs nut, 85
Dogs, poisoning of, 7, 48, 77, 85–86,
    147, 154, 156, 189, 214, 232
Donne, M., 53
Donworth, Ellen, 194–195
Dove case, symbiosis with Palmer
    case, 179
Dove, Harriet, 145, 173–176, 178–179
Dove, William, 173–180, 178–179,
    190
Doyle, Sir Arthur Conan, 125, 249
Drew, Samuel, 250–251
Dreyfus, Richard, xviii
Druggists *see* Pharmacists;
    Apothecaries
Du Bois Pacific Pills, 216
Duboscq, colorimetric method, 233

Duclos, M., 78
Dumas, Alexandre, 19, 57, 105–106,
    248–249
Dunstan, W.R., 30
Dupuytren (Guillaume), 163
Dyer, Dolores, 236
Dysentery, 7
Dyspepsia, strychnine for, 205

E

East India Company, 33–36
Easton's syrup, 203
Eau-de-Cologne, 90
Ecole de Médicine, 89
Edinburgh, 89, 122
*Edinburgh Medical and Surgical
    Journal*, 59
Edmunds, Christiana, 92, 186–192,
    240, 263
Egypt, 19
Eleanor of Aquitaine, 83
Electricity, xviii, 5, 68, 77
Electuarium de ovo, 26–27
Elegant synthesis, 225
Elephant bird, 30
Elizabeth I, Queen, 38, 84
Ellenborough, Chief Justice, 87
Emetics
    poisoning with, 105–107
    treatment for strychnine poisoning,
        16, 24, 53, 231
    usefulness of, 104–105
Emetic, tartar, *see* Antimony
    potassium tartrate
Encyclopaedia Britannica, 219
Encyclopaedia of Advertising Tins,
    216
Enghien, 100
England, *see also* Britain; London
    books on strychnine published in,
        69
    coroners in, 87
    herbals appear in, 23
    Mabel Jones returns to, 230
    middle, rage against, 245
    middle, something nasty in
        woodsheds of, 179
    poisons readily available in, 93, 183
    rarity of poisoning in, 84

regulation of poisons in, 92
Vaquier arrives in, 230
violent deaths in, 92
vomiting nuts available in, 24
Wainewright returns to, 116
Weakness of, the English 6
English, description of nux vomica in, 30
English, Hettie Wheeldon as, 242
English legal proceedings, low-key, 232
Epilepsy, 89, 137, 151, 155, 157–158, 193
Epps, Dr., 76
Erdman, Holger, 223
Ergot, 216
Erle, Mr. Justice, 142
Estrat, Rose, 92
Ethylene, 222
Evans, Spencer, 263
Examinations, 43
Expert witnesses, 87, 142, 141
Exhetwick, 148
Extortion schemes, *see* Blackmail

F

Fabia Amara, 214
False Angustura, 58–60
Faculty of Medicine, Society of, 1–3, 7
Faraday, Michael, 81, 169
Faria, Abbé, 247
Farr, William, 86
Faulkner, Julia, 193
Fellows' Syrup, 203
Fenchurch Street, London, warehouse, 36
Fertility, 78
Fevers, 5
First World War, 223–224, 239, 241–242, 245, 253
Fishberry, 80
Fisher, Ishmael, 129
Fisher, John, 84
Fitzgerald, John H., 262
Flambeau, Hercule, 253
Fleischmann's Hotel, Leeds, 175
Fletcher, George, 167
Florence, 23
Fluckiger, 29

Foderé, F.E., 89, 92
Foersch, explorer, xvii
Food
  adulteration of, 211
  contamination of, by strychnine, 210
  strychnine as, 72
Forebrain, 12
Forensic science/analysis, 89–92, 106–108, 139, 141–160, 168–173, 170, 176, 181, 186, 188, 195–197, 229, 233, 237, 243, 250, 253, 256–257, 263–264
Forgery
  capital penalty for, 117, 182
  Palmer's, 129, 159
  Wainewright's, 113, 117
  Wakley, of pamphlet by, 168
Formulaire Pour la Preparation et l'Emploi de Plusieurs Nouveaux Medicaments, 67
Forsyte Saga, 126
Foster, bus conductor, 157
Fougnies, Gustave, 90
Fountain, Dennis, 263
Fouquier, Pierre-Eloi, 2–13, 50, 67–69, 73, 76, 247–248
Fouquier-Tinville, 2
Fourcroy (Antoine François), 47
Foxes, poisoning of, 210
France, *see also* Paris
  centralised country, 76
  devastation of, 239
  disappearance of strychnine poisoning in, 91
  expeditions launched from, 48
  Lloyd George goes to, 244
  Louis XVIII restored to throne of, 247
  manufacture of strychnine in, 213
  many scientists in, 2
  medical opinion in, 76
  peasants in, 6
  poisoning widespread in, 98
  records of poisoning in, 91
  regulation of poisons in, 11, 77, 98–99
  scientific climate in, 18
  use of pesticides in, 210
  Vaquier arrives from, 230–231

Wainewright travels to, 115–116
weakness of, the French, 6
Fredoy, Marie, 9
Free Pentecostal Holiness Church, 262
Freet, Mary, 165, 257
French lunar pills, 215–216
Frere, Mr., 138
Frogs, poisoning of, 147, 195–196
Fruit, 6
Fuckoffsky, 260

G

Gainsford, 184
Galen, 3, 4, 20, 40, 54
Galenicals, 40
Galipea officinalis, 58
Gallus, physician, 26–27
Gamekeepers, 186, 210, 256, 265
Gamekeeper, The, 210
Garbler of spices, 37–38
Garcia, William R., 202
Gardiner, Mr., 136, 139, 147
Garrett, Isaac, 188–189
Garth, Samuel, 33
Gastrointestinal tract, *see* Alimentary
canal
Gattermann's Practical Methods, 234
Gay, Dr. Robert, 157
Gedney, 184
Gelatine capsules, 194
Gelsemine, 255
General Medical Council, 43
Gentian, 27, 214
Geoghegan, Mr., 216–217
George V, King, 241
Gerard, John, 25–26, 33, 38, 85
Germans, 6, 41, 241
Germany
adulteration of beer in, 213–214
morphine isolated in, 49
nux vomica name in, 29
physicians in, 7
Wheeldons' connection with, 242
Gibson, Dr., 191
Giesing, Dr., 260
Glasgow Royal Infirmary, 145
Gloire d'Anvers (ship), 35
Glycine, 12–13
Goa, 22

God
visitation of, death by, 88
works of, 152, 187
Goldenseal, 251
Goldsmith, Oliver, 42
Gomez, Leonor, 21
Gophers, 229, 265
Gordon, secret agent, 242, 244
Goruckpore, 207
Graham, Daniel, 39
Graham, Dr., 115
Graham, Thomas, 211, 213
Granville, Dr. Augustus, 76, 81
Granules, spinal, 155, 157
Grattan, Harriet, 114
Graves, Robert, 134
Grawitz, Dr., 259
Gregorius, Hagar, 249
Griffiths, George, 112–113, 114, 119,
123
Griffiths, Ralph, 112–113
Grocers' Company, 37–38
Gum-gutta, 92
Guwo Upas, xvii
Guy's Hospital London, 146, 243

H

Habituating poisons, 17, 67
Hahnemann, (Samuel), 207
Haines, Professor Walter S., 193
Hall, Marshall, 178, 199
Halloween, 259
Haly Abbas, 20
Hamburg, 58, 264
Hamilton, Mr., 114
Hammond, Dr. William, 54
Hampshire Independent, 63
Handbury, 29
Handschius, Georg, 26
Handcocks, Sarah, 114–115, 118
Hanging
of Cream, 198
of Dove, 179
of Elizabeth Pearson, 184
of Ethel Major, 257
of George Horton, 185
of Mary-Ann Britland, 185
of miscreants, 182
of Palmer, 161

public, 93
of Roderigo Lopez, 84
statistics, 93, 177
Hanssen, chemist, 225
Harland, Dr., 137–138
Harmsworth, Alfred, 246
Harris, Dr. W.T., 13–17
Harrison, Henry, 173–176, 179
Harrison, John, 262
Harvard, 235
Harvey, G.F. Company, 194
Harvey, Lou, 196
Hawkins, Dr., 76
Hawkins, druggist, 134
Hawkins, Mr. Justice Henry, 196,
    216–217
Haworth, W.N., 238
Hazlitt, William, 121
Headache, 67
Heart, treatment with strychnine, 79
Heavy horsemen, 35
Hednesford, 136
Heinrich, Dr. Edward, 229
Hellebore, 216
Hemiplegia, 11, 68, 75
Hemsley, George, 262
Henbane, 92
Henderson, Arthur, 240–241
Henry II, King of England, 83
Henry VIII, King of England, 84, 257
Herbals, 23
Herepath, William, 62, 156, 176
Hernia, 67
Heroin, 204
Herron Pharmaceuticals, 263
Hervey, Charles, 84
Hicks, Thomas J., 202
Hickson, Caroline, 64
Higden, Ranulf, 83
Hieronyma Spara, 84
Hindustan Times, 206
Hitler, Adolf, 259–260
Hobart, 117
Hodson, Mary, 184
Hodson, Palmer's mother's lover, 126
Hofmann, Albert, 262
Hofmann, August Wilhelm, 211–213
Hogarth, William, 39
Holland, 59
Holmes, Annie, 254

Holmes, Sherlock, 249
Hotel Russell, London, 230
Hotel Victoria, Biarritz, 229
Holt, Frederick, 178
Homeopathy, 206–207
Hooghly River, 33
Hôpital de la Charité, 2, 7, 9–12, 247
Hops, 212, 214
Horncastle, 255
Horoscopes, 173–174
Horses, doping of, 154
Horsford, Walter, 210, 254
Horton, George and Kate, 185
Houghton, Dr. J.H., 77
Household poisoning, 92
House of Lords, 177, 183
How Does Your Garden Grow,
    253–254
How to Murder Your Wife, 181
Hull, 257
Humane Society, 75
Humours, four, 4
Hungary, 58
Hunted Down, 116
Hurst, E.L., 238
Husayn, 20
Husemann, 59
Hydrargyrum con creta, 44
Hydrastine, 251
Hydrastis Canadensis, 251
Hysteria, 64, 69, 78, 175
    about strychnine use, 77

I

ibn Sina, 20
ibn Wahsiyyah, xix
Idiopathic tetanus, 143–144, 148, 151,
    158
Illinois, 193, 198, 235
Illustrated London News, 245
Illustrated Times, 146, 162, 168–169
Imponderable quantities, 147
Impotence, 78, 208
Incontinence, 8
Independent World Workers, 242
India
    abortion in, 215
    Chakravarti arrives from, 223
    control of lions and tigers in, 210

export of drugs from, 30, 209, 264–265
nux vomica aphrodisiac in, 207
occurrence of strychnine and brucine in, xix
plants from, 7, 28–29, 33
poisoning in, 84–85
prohibition of British subjects from, 34
snakewood in, 22
tetanus in, 148
traditional medicine in, 66–67
weightlifters of, 206
India pale ale, 211–212
Indian cockle, 80
Indian leaf, 27
Industrial societies, diseases of, 6
Infertility, 78
Inglethorp, Mrs., 252
Injections, strychnine, 80
Inland Revenue, Board of, 214
Inquests, see Coroners' inquests
Inquisition, Spanish, 21
Insanity
    alienists and, 178–179
    law on, 76–177, 257
    of Cream, 195, 197–198
    of Dove, 173, 176–179
    of Edmunds, 189–192
Instrumental methods, 234, 237
Intemperance, 5
Intestines, see Alimentary canal
Intestinal worms, 67
Ipecacuanha, 77, 262
Iraq, 20
Ireland, 241–242
Irritant poisons, 71
Irving, David, 259
Isaac of Salerno, 20
Isabella, Queen, 21
Isotopic labelling, 265
Italy
    poisoning in, 84, 98
    transit of drugs through, 19–20

J

Jackson, chemist, 214
Jack the Ripper, 197–198, 255
Jalap, 92

James I, King, 38
James, Dr. Robert, 42, 104
James's Powders, 42, 104
Janhin, Francois, 9
Janus Weathercock, 112
Java, xvii–xviii
Jaws, xviii
Jems, 208
Jesuit bark, see Cinchona
Jesuits, 28, 165
Jews
    expulsion of from Spain, 21
    Portugese, 84
    translators, 20
J.H.G. of Bath, 209–210
Jockey Club, 182
John Bull, 59, 213
Joliet Prison, 193, 198
Jones, Alfred, 230–232
Jones, Dr. William, 133, 135, 154
Jones, Mabel Theresa, 229–233
Jones, Mr., pharmacist, 64–65
Journal of the American Chemical Society, 236
Juries
    coroners', 65, 86
    in Palmer case, 150, 153, 160
    in Wainewright case, 116
Jussieu (Antoine Laurent), 48

K

Karbala, 20
Katherine of Aragon, 84
Kentucky, 250, 262
Khubz al Ghurab, 20, 23
Kidneys, 171
Kirby, John George, 93
Kirkby-on-Bain, 255–256
Koester, Dr., 259
Korte, Friedhelm, 264
Krahenaugen, 29
Kuchila, 29, 34–35, 67, 84
Kuchila molung, 30

L

Labrador, steamship, 194
Laburnum, 170

Lallemand, Claude François, 81
Lambeth, 192
Lamson, George, 93, 95, 196
Lancet, The, 67, 78, 88, 146, 169, 171, 197, 209
Langley, Mr. Baxter, 161
Lard, 54
Latvia, 58
Laudanum, *see* Morphine
Laune, Gideon de, 38
Laurel, 92
Lavoisier, Antoine, 1–2, 47, 50
Laxative, strychnine as, 204–205
Lead poisoning, 17, 67–68, 92, 211
Lebret, M., 100–101
Lee, Major, 242
Leeching, 4, 9, 102
Leeds, 173–175
Legal medicine, 89
Lemmon, Jack, 181
Lenitive electuary, 44
Leschenault, Jean Baptiste, 48–49, 51
Letheby, Dr. Henry, 142, 156, 163, 168, 172, 188
Leuchs, Hermann, 225–226
Leu-sung-kwo, 28
Liber Regius, 20
Liebig, Justus von, 42, 83, 171, 212, 217, 248
Life insurance, 88–89, 113, 116, 122, 126–128, 169, 183, 185, 264
Light horsemen, 35
Lignum colubrinum, 21, 23, 61
Lincolnshire, 255
Linden House, 112–113, 115–116, 119
Linnaeus, Carl von, 28–29, 41, 51, 55, 223
Lions, poisoning of, 210
Lirriper, Lady, 249
Lisbon, 23
Little Dorrit, 117
Liver
    detection of strychnine in, of dog, 156
    disease as "nux vomica, strychina," 207
    importance of, 4
    metabolism of drugs by, 74–75
    treatment of with strychnine, 78
Liverpool College of Chemistry, 42

Liverpool University, 224
Lloyd George, David, 239–245
Lloyd, John Uri, 250–251, 259
Lloyd Library, 251
Lock-and-key mechanism, 74
Lockjaw, *see* Tetanus
Locock, Dr., 80, 114–115, 118
Loline, 266
London
    Accidental poisoning in, 65–66
    Apparent suicide of a young woman in, 88
    apothecaries, 38, 40–41
    chemists and druggists outside, 40
    Cream visits, 192, 194–195
    Docks, 35–36
    druggists' shops in, 43
    drug warehouses, 36–37, 59
    Elizabeth Mills moves to, 163
    Palmer visits, 154
    trades of, 41
    Vaquier visits, 230
London Hospital, 143
London Pharmacopoeia, 69
London University, 224
Lopez, Roderigo, 84
Lord Chancellor, 183
Lors, Fred, 202
Loudon, Mr., xvii
Louis XIV, 104
Louis XVIII, 247
Low, Mr. Justice, 244
LSD, 261–262
LSD: My Problem Child, 262
Luzon beans, 28
Lymphatic system, 5–8, 11
Lytton, Edward Bulwer, 123–124

M

Mackean brothers, 121
Mackenzie, Sir James, 79
Madras, 30, 84, 209
Magellan, 27
Magendie, François, 2, 48, 53–55, 67, 76, 75, 92, 105–106, 164, 172
Magistrates, Kent County, 87
Magnetism, animal, 2
Magnolia, 58
Mair, Dr., 84

Maisch, J.M., 32
Maissemy, 2
Major, Arthur and Ethel, 210, 256–257
Malingerers, 10–12
Malaise, 22
Malaria, 3, 27, 57
Malassis, M., 101–102
Malaya, 34
Malt, 214
Manchester University, 222–224
Mandovy, River, 23
Manual for Students Preparing for Examination at Apothecaries Hall, 44, 142
Maqianzi, 264
Marathon, Olympic, 201–202
Marconi Company, 231
Marco Polo, 19, 21
Mar, Earl of, 122
Markov, Georgi, 243
Marsden's vermin killer, 184
Marsh, Alice, 195
Mart, Dr., 75
Martell's French female pills, 216
Martignon, M. et Mme., 100–101
Martin, Mr., 193
Masefield, John, 36
Mason, Alfred and Winnie, 240–242, 245
Massachusetts, 202
Massachusetts Institute of Technology, 234
Mass spectrometry, 237, 251, 263
Masturbation, 6
Materia Medica, 175
Mathers, Helen, 249
Matrons, panel of, 191
Matthiolus, 23–24, 72
Maugham, Somerset, 192
Maximilian I, Emperor, 26
May, Adam, 188
Mayhew, Augustus and Henry, 169
Maynard's sweet shop, 187–189
McCulloch, J.W., 194, 216 – 217
McDonald, Alexander, 242
McDonnell, Dr. William, 157
McGill University, 192
Meat, 6
Media, handling by Taylor, 169

Medical and Physical Society of Calcutta, 66
Medical jurisprudence, 89
Medical profession, *see also* Doctors
    adopts strychnine, 67
    foundation of modern, 44
    no bulwark against disaster, 65
Medical Register/Directory, 93–94
Medical Registration Bill, 131
Medical Times, The, 212
Medicine
    Arab, 20
    Central European, 26–27
    control of, struggle for, 38–39, 43
    French Academy of, 1
    herbal, 27–28, 47–49, 206
    holistic, 2
    homeopathic, 206
    oriental, 21–22
    Palmer studies, 127
    plants useful in, 3–4
    powerful boost to, in France, 1
    state of in early nineteenth century, 3
    shortcomings of Victorian, 144
    tartar emetic, use in, 104
Medico-Botanical Society, 76
Medulla, stimulation of by strychnine, 205
Melting points, 52, 60
Mercury, mercurials, 4, 44, 104, 129, 131, 186
Merré, Joseph, 11
Merritt, Ann, 165
Mesmerism, 77, 249
Metabolic disorder, 4
Metabolism, of strychnine, 73, 75, 170–172
"Meta-strychnine", 260
Metatone, 81, 204, 218
Methel, 23, 25
Metropole Hotel, London, 194
Mexico, 25
de Meyrignac, H.P., 54
Mice, poisoning of, 184, 211, 217, 265
Middle Ages, 3, 7, 26, 248
Middle England, 179, 245
Middlemarch, 199
Middlesex Street, London, 36
Middleton, Lord, 25

Miliary tuberculosis, 164
Millar, Bruce, 232–233
Mill City, Arizona, 262
Miller, Charles, 187
Millipedes, 26
Mills, Elizabeth, 131–133, 151, 163, 179
Miscarriage, 215
Missouri, 202
MIT, see Massachusetts Institute of Technology
Mithridates, 40
M'Naghten rules, 177–179
Molasses, 103
Molecular formula, 222
Molecular Moloch, strychnine as, 15
Moles, poisoning of, 265
Molly Strychnine, 260
Mongoose, 22
Monkeys, 30
Monks, Catalonian, 57
Montreal, 194
Moore, Mr., 77, 145
Morley, Mr., 142, 175–176
Morphine/opium/laudanum
    addiction to, 121–123, 194
    antidote to strychnine, 16, 53
    beetles consuming, 267
    bitterness of, 104, 107
    in colour tests for strychnine, 251
    forensic documentation of, 92
    isolation of, 49, 51, 106
    medicinal use of, 64, 131
    poisoning with, 85, 105–108, 112, 118, 132, 134, 160, 162, 255
    security of packing, 37
    structure of, 225
    in tonics, 204
    untraceability of, 90
Morris, Mr., 214
Morson, Thomas, 67
Moscow, 10
Muff, Annie, 230
Mules, 25
Müller, Dr., 186
Muller, Eudoxia, 236
Mulliner, Cyril, 254
Mundie, Dr. G., 203
Munich, University of, 224
Munster, 261

Murchison, Terence D., 261
Murder Considered as one of the Fine Arts, 120–121
Murray, W.H., 194
Muscular people, diseases of, 5
Mushroom, poison, 186
Musk, 35
Muspratt, Sheridan, 46
Myatt, George, 129–130
Myatt, John, 139
Mynheers, 59
Mynsicht, Adrien de, 104
Mysterious Affair at Styles, The, 247, 252–253

N

Napoleon, 10, 111, 124
Napoleon III, 161
National Drug Company, 204
Nature, as sex organ, 208
Naturopaths, 206
Neill, Thomas, see Cream
Nerve gas, 259, 268
Nerves
    impulses, study using strychnine, 264
    injury to, 4
    motor, 5
    nervous system, 6, 7
Nervines, 71
Nervous indigestion, 75
Netherlands, 59
Nettle, racehorse, 127
Neuralgia 69, 75
Neurological research tool, strychnine as, 205
Neurology, 5
New Cross Inn, Leeds, 174
Newgate Prison, 117–118
New Herball, 23
Newspapers
    amusement from, 267
    complaints in, about strychnine restrictions, 264
    Dove case and, 179
    growth of, 167–168
Newton, Charles, 134, 137, 154, 158–159

New York, 202, 216
Nichol, Dr. J., 79
Nicolaus of Salerno, 20
Nicolaus Praepositus, 20, 26
Nicotine
    antidote to strychnine, 16–17, 54,
        56
    poisoning with, 90, 253
    untraceability of, 90
Nitric acid, 52, 62
Nitrogen Nero, strychnine as, 15
Nizamut reports, 207
Nobel Prize, 224, 236–238
Noirtier, M., 248
Nonketonic hypoglycaemia, strychnine
        as research tool for, 205
Norris, James F., 234
North Carolina, 262
Northcliffe, Lord, 245
Nottingham, 111
Nucis vomicae, 20
Nunneley, Dr. Thomas, 155, 158, 160,
        172
Nux vomica
    abortifacient, 216
    adulteration of, 69–70
    adulteration of beer with, 214
    alcoholic extract of, 11
    antigas pills containing, 260
    aphrodisiac, 208
    bitterness of, 4
    confusion with Datura, 24–25
    constipation treated with, 77
    dangers of, 72–73
    delirium caused by, 12
    demon, labelled as, 213
    Desportes' thesis on, 49
    dispensing of, 65–66
    domestic use, 39
    enemas, 9, 12
    Fouquier uses, 7–13
    German doctors, use by, 7
    homeopathic use, 206
    Indian medicine, use in, 66–67, 148
    Indian poisoners, use by, 84
    insanity caused by, 12
    isolation of strychnine from, 50–51,
        60, 143
    insomnia caused by, 12
    misnaming of, 24

    reaches the West, 19–20, 23
    ready availability of, 183
    rise in popularity, 67
    roasting of, 67
    suicide by, 88
    tonics containing, 203–204
    trade in, 209–210
    variation in strength of, 76
    warehousing of, 35–37

O

Occult sciences, 248
Oenanthe Crocata, 170
Offences against the Person Act, 241
Old Bailey, 93, 143, 177, 192, 195,
        241, 244–245, 252
Ollier, Mr., 262
Old Indian Medicine Co., 204
Original Compound Wa-hoo Bitters,
        204
Olympiad, third, 201
Ontario, 192
Operations, use of strychnine in, 79
Opisthotonos, 12, 16
Opium, see Morphine
Orfila, Mathieu, 89–90, 106, 119, 145,
        164, 172, 262
Orwell, George, 239
O'Shaughnessy, Sir William Brooke,
        61–62
Otto, Julius, 91, 164
Overdose, accidental, of strychnine,
        14–17
Overexcitement, 157–158
Oxford University, 223–226, 238

P

Padwick, Mr., 128, 140, 142
Paedophilia, 249
Pain
    strychnine as research tool for, 205
    and pleasure, sensation, 5
Palladium Insurance Company, 113
Pale Ale, India, 211–212
Pall Mall Gazette, 191
Palmer Act, 143
Palmer, Annie, see Brookes

Palmer case
  progress since, 196–197, 233
  public hysteria over, 98
Palmer, George, 164
Palmer, Joseph, 126
Palmer, Sarah, 126–128, 158, 162
Palmer, Walter, 127–128, 131, 140,
    158–159, 161–162, 164, 169
Palmer, William
  appearance in fiction, 252
  cause celebre, 99, 182
  children, 162
  debts, 127–128
  family history, 125–126
  financial problems, 150
  incompetence of, 172
  guilt or innocence of, 160
  marriage, 127
  murders Cook, 128–139
  murders wife, 127
  parvenu, 182
  residence, 263
  studies as apothecary, 44, 126–127
  trial of, 13, 62, 73, 75, 88–89, 103,
    107, 141–163, 167, 170
  wickedness of, 173
  wit of, 149, 161
Palmer, William, Junior, 162
Palmerston, Lord, 125
Pampe, Julius, 87
Panadol, 263
Panic, public, about strychnine, 180
Pankhurst, Emily, 245
Pantegni, 20
Paracelsus, 39, 71
Paracetamol, 263
Paralysies saturnines 67
Paralysis, 8–13, 68, 75
Paraplegia, 68, 75
Paris, see also France
  Alkaloid research in, 50–55, 59–61
  Castaing poisoning in, 99–102, 106
  Christison studies in, 89
  French Academy in, 1
  hospitals in, 2
  ordering strychnine from, 214
  Taylor studies in, 146
  use of strychnine in, 75, 105
  Wainewright in, 115
  Westwood studies in, 43

Paris, Charles Martin, 8, 10, 11
Parke Davis Co., 204
Parker, Inspector Edward, 41
Parry, Serjeant, 189–190
Partridge, Dr. Richard, 157
Pasteur, Louis, 4, 144
Pausanias, 54
Payen, Dr., 213
Payphones, 260–261
Pearson, Elizabeth, 184
Peasants, French, 6
Peel, Sir Robert, 177
Pelletan, Professor, 102
Pelletier, Bertrand, 50
Pelletier, Pierre-Joseph, 50–53, 55,
    59–62, 67, 92, 119, 164, 213,
    221–222, 228
Penicillin, 74, 226
Penis, Cook's, 129, 155
Pennington Street, London, 36–37
Pennyroyal, 216
Pepita, 28
Pepper, 129, 267
Perfumes, denaturation of, 217
Period pains, 69
Perkin, W.H., Jr., 223–225
Petticoat Lane, London, 36
Pharmaceutical Journal, 99, 253
Pharmaceutical Society, 44, 205, 265
Pharmacopoeias, xviii, 23, 205, 208
Pharmacists/Druggists, see also
    Apothecaries; Chemists
  Abortifacients sale by, 216
  Agatha Christie and, 252–253
  Alfred Mason lectures in, 240
  Castaing visits, 101
  Christiana Edmunds deceives,
    188–189
  Cream accuses, 193
  Ethics of, 35
  French, regulation of, 99
  Indistinguishable from chemists,
    41
  Laxity of, 93–94, 98–99, 183
  Mary Dixon visits, 185
  Mistakes by, 65–66
  Rugeley, 134, 154
  Strychnine/nux vomica availability
    from, 88, 93, 210, 213, 229
  Strychnine migrates to, 203

M. Touery, swallows strychnine, 54
Unregulated, 40, 43–44
Vaquier buys from, 230–231, 233
Pharmacy Act, 1868, 92
Philippines, 27
Philadelphia, 261
Phillips, George, 214
Phlegmacies, 5
Phoenix, Arizona, 262
Phosphorus, 91
Phrenology, 161
Physicians, *see* doctors
Physiology, 1–2, 4
Piddington, Mr., 61
Pidoux, H., 68, 74, 207–208
Pigache, Dr., 102
Pigny, Guillaume, 10
Pillcock, Paracelsus, MD, 65
Piper, S.E., 211
Plasters, blistering, 8, 9
Platinum, 172
Plenck, J.J., 89
Pleurisy, 100
Poachers, 210
Pocket Full of Rye, A, 253
Poirot, Hercule, 247, 253
Poisoned arrows, 245
Poison nut, 29
Polestar, racehorse, 128–129
Polymorphism, 52
Polyzonium rosalbum, 266
Ponsin, Mme., 9
Pornography, Cream's, 194, 197–198
Porta, Giovanne, 85–86
Portland, Duke of, 182
Port of London Authority, 36
Post-mortem/Autopsy
    on Alfred Jones, 230, 233
    on Annie and Walter Palmer,
        161–162
    on Arthur Major, 256
    on Mr. Appleton, 253
    on Ballet brothers, 100, 103,
        106–107
    on Mrs. Bowles and her son, 186
    on Carroll Rablen, 229
    on Cook, 136–139, 146–147, 157
    on Daniel Stott, 193
    on Fougnies, 90
    on Helen Abercromby, 115, 118
    on Matilda Clover, 195
    on Mrs. Dove, 176
    reluctance to carry out, 87–88
Potassium bromide, 252
Potassium dichromate, 91
Potency, sexual, 78
Pott's disease, 68
Powder of post, 219
*Practical Methods of Organic
    Chemistry*, 234
Prager, Lyon, 34
Prague, 26
Prairie Blossom pills, 216
Pratt, Thomas, 126–129, 140, 150
Pregnancy, *see also* Abortion
    Christina Edmunds, 191
Prelog, Vladimir, 227
Prescriptions, illegible, 65
Prodrugs, 74
Professional ethics, physicians', 80
Professional qualifications, 43
Prolapsed rectum, 67
Prudential Assurance, 185
Prussian blue, 184
Prussic acid, 162
Pseudo-spermatorrhoca, 78
Psychiatry/Psychology, 178–179
Publicans, 214
Pullman, Irja, 236
Pulvis nucis vomicae, 66
Pulv. Strychnos, 66
Punch, 42, 65, 98
Purging, 4

Q

Quack medicine, 42–43
Quaker Buttons, 29
Qualifications, professional, 43
Quassia, 212, 214
Quebec, 194
Quil, 22
Quinine, 3, 50, 58, 109, 235
Quirpile, 22

R

Rabbits, poisoning of, 53, 146
Rabies, 67

Rablen, Carroll and Eva, 228–229
Rademacher, Johann, 207
Radioimmunoassay, 237
Rainforest, 267
Railways, 136, 150, 178
Rambach, Johann Jakob, 58, 60
Rats, poisoning of, 184, 217, 265
Raven Hotel, Shrewsbury, 129–130, 168
Ray, John, 28
Rearrangements, of carbon skeletons, 226
Receptors, nerve, 12
Red Sea, 19
Rees, Dr., 146
Reeves, Mrs. Henry, 249
Reflexes, nervous, 12
Reid, Mr., 130
Remer, W.H.G., 92
Resin, pills containing, 148
Restlessness, caused by strychnine, 14
Revolution, French, 1, 6
Rheumatism, 67
Rhubarb pills, 131
Richardson, Dr. B.W., 158
Ricin, 243
Ricinus communis, 170
Risus sardonicus, 12, 135, 202
De Rivera, Dona Francisa, 3
Rizi, Mr., 244
Roasting, of nux vomica, 67
Roberts, Charles, 134
Robinson, Lady Gertrude, 224
Robinson, Sir Robert, 222–228, 238, 265
Rochester, Bishop of, 83
Rohuna, 61
Rohun bark, 61
Roland, Dr. R., 69
Romsey, 63
Roose, Richard, 83–84
Roosevelt, Theodore, 201
Rousseau, M., 8, 12
Roy, M., 92
Royal College of Surgeon,s 45
Rugeley, 55, 125–127, 130–139, 142–143, 147, 154, 156, 158, 161, 168, 170, 179, 215, 248, 252, 263, 265
Russia, 58

Rutnagherry, 34
Ryley, Dr. Beresford, 191

S

Sabine, 92
Sadness, weakness due to, 6
Sale of Poisons Act, 188
Saint Andrews University, 224
Saint Bartholemew's Hospital, 84
Saint Cloud, 101–102
Saint George's Hospital, London, 76
Saint Ignatius beans, 28, 50–51, 106, 148, 213
Saint Ives, 210
Saint Louis, Missouri, 201
Saint-Meran, Madame de, 248
Saint Neots, 210
Saint Petersburg, 58
Saint Thomas's Hospital, London, 94, 192, 195
Salamanca, 22
Salerno, 19–20
Salicine, 64
Salpetrière Group, 249
Salt, Mr,. 134
Salts, of strychnine, 52
San Francisco, 202
Sap, 48
Scammony resin, 92
Scheele (Carl Wilhelm), 49
Scepticism, about strychnine, 75, 79
Schnoll, S.H., 261
Scilla peruviana, 55
Scrofula, 8
Scuffle hunters, 35
Secret agents, British, 242, 244
Senna, 77
Sennett, Agnes, 145
Sensory perception, strychnine effect on, 207
Serapio, 24–25
Sergison Smith, Mrs., 63, 64–65, 145, 156, 252
Serteurner, (Friedrich Wilhelm), 49
Sex, fears of, 6
Sexual performance, strychnine effect on, 207–208
Shadwell, London, 36

Shee, Serjeant, 142, 144, 149–155, 162, 170
Shell-shock, 179
Sheriff
  court, 191
  U.S., 229
Sherwood Foresters, 242
Shipping, Ministry of, 240
Shock, surgical, strychnine treatment for, 205
Shock, weakness due to, 6
Short, F.W., 30
Shrewsbury, 129–130, 147, 154, 168
Shrivell, Emma, 195
Shurab, 207
Sibpur, 33
Sidney, Mr. Alderman, 125
Sidney, Violet, 81
Sigmond, Dr., xix, 76–77
Simmons, Mr., 156
Simples, 40
Sin of Hagar, The, 249
Sirius, racehorse, 129
Skin absorption, of poisons, 261
Smethurst, Dr., 163
Smith, Adam, 39
Smith, F.E., 242
Smith, Jeremiah, 131, 158–159
SmithKline Beecham, 263
Smith, Mrs Sergison, 63, 64–65, 145, 156
Smoking, see Nicotine
Snakes, 22, 66, 205, 211, 267, 263
Snakewood, 21–22, 61, 148
Society of Oddfellows, 185
Society of Public Analysts, 211
Soil, absorption of strychnine by, 248
Solanaceae, 28
South Africa, 13
South America, 3, 57
Southampton, 240, 242–243
Soymida febrifuga, 61
Spalding, 184
Spasms, muscular, 12
Spectator, The, 192
Spermatorrhoea, 78
Spiders, 267
Spilsbury, Bernard, 233, 244
Spinal column, 5, 12, 54–55, 157
Sponges, 267

Squyer, Edward, 84, 268
Stability, chemical, of strychnine, 75, 225–226
Stack, Ellen, 193
Stafford Rural Constabulary, 128
Stafford/Staffordshire, 130, 139, 142–143, 149, 153, 160–161, 164, 168, 182, 210
Stamford Street, London, 195
Stas, Jean Servais, 90–91, 253
Steam boilers, 72
Steggall, John, 44–45, 127, 142, 154
Stegobium paniceum, 267
Stethoscope, 191
Stevenson, Dr., 184, 186, 195–197, 216–217
Stevens, William, 135–137, 139, 153–154, 163, 169
Stolen Strychnine, 254
Stomach, see Alimentary canal
Stomach contents
  Cook's, 137–138, 146–147
  Helen Abercromby's, 115
Stott, Daniel and Julia, 193, 228–229
Stramonium, 85
Strangulated hernia, 67
Straubenzee, Nadège, 249
Strendenberg, Mr., xvii
Strength, muscular, 6–7
Stringtown on the Pike, 250, 259
Stroke, 55
Strongman, Tina, 260–261
Structure, chemical, of strychnine, 226–228
Strychnine Arms, The, 143
Strychnine in the Soup, 254
Strychnine, racehorse, 173, 180
Strychnos
  colubrine, 31, 61
  ignati, 28
  nux-blanda, 30
  nux-vomica, xix, 28–29, 61
  potatorun, 29
  tieute, xix
Submarine warfare, 239–240
Suffocation, 12, 16
Suffragettes, 243, 245
Suez Canal, 27
Suicide, by strychnine, 88, 262
Sulphuric acid, 52, 62

Sunstroke, 103
Sulusuquir, xix
Surgeons, Royal College of, 45
Surgeons, use of strychnine by, 79
Surgical shock, strychnine treatment
    for, 205
Surrey, 186, 230–231
Swanwick, 185
Sutton-under-Brailes, 262
Swellings, scrofular, 8
Sydney, 224
Sykes, Sir Mark, 124
Synthesis
    Robertson's, 224–225
    Woodward's, 235–237, 264
Syphilis, 129, 137, 157–158, 179, 195

T

Tafel, Julius, 225
Talbot Arms, Inn, 130–131, 135–139,
    263
Tartar emetic, see Antimony
    potassium tartrate
Tasmania, 117
Taxonomy, 85
Taylor, Alfred Swaine, 63
    career of, 145–146
    Cook's inquest, attends, 146
    Cook's stomach and, 147, 150
    criticism of, 151–152, 156
    engaged by Stevens, 135, 139
    failure to find arsenic, 163
    failure to find strychnine, 139, 141,
        163, 170, 173
    false press reports and, 168
    instant book by, 169
    life insurance and, 89
    morphine and, 107–108
    poisons rabbits, 150
    reluctance to name poisons, 95
    strychnine detection and, 146–147,
        156
    theory of perfect absorption,
        156–157, 168–172
Taylor, Mr. Francis, 64–65
Teak, 34
Telescope, 191
Temperament, 5
Tennessee, 262–263

Tetanine, 55
Tetanospasmin, 13
Tetanus, 13, 55, 133, 141, 143–144,
    148, 151, 158
Tharm, Eliza, 128, 155
Theory of perfect absorption, 168–172
Thin-layer chromatography, see also
    TLC, 237, 261
Thirlby, Mr., 134
Thorn apple, 23, 25, 84
Thornton, Mary, 127, 162
Thuggees, 23, 84
Tic douloureux, 75
Tigers, poisoning of, 210
Times, The
    correspondence about forensic
        evidence, 168
    dominates newspaper market, 167
    editorial on Dove case, 178
    Harrison reads from, 174
    letter from Garrett, 189
    letter from Mayhew, 169
    modern complex, 37
    reports Edmunds trial, 191
    reports Horsford trial, 210
    reports Olympics, 201
    reports Palmer hanging, 16
    role in fermenting public panic,
        179–180
Tittletown, California, 229
Timor, 48
TLC, see also Thin-layer
    chromatography, 237, 261
Tobacco, see also Nicotine
    Bocarmé grows, 90
Todd, Dr. Robert, 144, 153
Tonics, strychnine, 14, 203–205
Tooting, 195
Toronto, 264
Torquay, 253
Tortoiseshell, 23
Toucans, 30
Touery, Pierre-Fleurus, 54, 56
Toxicology, Victorian ignorance of,
    163
Traumatic tetanus, 143–144
Treacle, 26–27
Treatise on Poisons, A, Christison's,
    45, 71, 119
Treatise on Therapeutics, 68

Trelawney, Blake, 249
Trinidad, 57
Trismus, 13
Tropinone, 224
Trppyn, Alice, 83
Trousseau, Armand, 68–70, 74–75, 207–208
Tschettik, 164
Tsuan-Chow, 19
Tuberculosis, 100, 164
Turmeric, 34
Turner, William, 23–24, 26
Turnham Green, London, 112
Twort, Dr., 186

U

U-boat warfare, 240
Ulcers, 24
Undersherriffs, 191
United States
    abortion in, 215–216
    court judgements, 204, 208
    Cream in, 192–193
    Evidence about Cream's sanity from, 197
    First World War and, 239
    Nux vomica name in, 29
    Olympics in, 201–202
    Patent medicines in, 204
Universal Life Assurance Company, 159
University College, Southampton, 240
Upas, xvii–xix, 49, 164
Urban myths, 261
Ure, Dr., 120
Urinary incontinence, 77
Urinary retention, 68
Urine sampling, athletes, 205–206
Uxbridge, Lord, 126

V

Vagal centre, stimulation of by strychnine, 205
Valium, 54
Van der Warker, Ely, 216
Vanhove, M., 8, 9, 12
Vaquier, Jean Pierre, 229–234

Vasco da Gama, 21
Vasomotor centre, stimulation of by strychnine, 205
Vaughan, Mr. Justice, 117
Vaults of Erowid, 261
Vauquelin, Nicholas, 47–48, 50, 55, 106–107, 164, 172
Vegetable poisons, undetectability of, 90, 103, 146, 148, 196
Vegetables, 6
Venereal disease, 42, 153
Venezuela, 57–58
Venice, 19, 27
Venice treacle, 27
Viagra, 208
Victoria and Albert Museum, London, 36
Victorian age and society, 181–182, 184, 186, 210
Victorian science, limitations of, 170
Victorians, famous, 80
Vienna, 26, 58
Villefort, Madame de, 248
Vinegar, 90
Vining, John, 164
Virginia, 262
Viscum monoicum, 30
Vitamin C, 238
Vivisection, 54, 126
Vogel, W.H., 261
Vomiting
    Cook's, 131–133
    Elizabeth Mills's, 131
    exacerbates disease, 105
    morphine causes, 104
    Nux Methel and Nux Vomica, cause of, 25
    primitive medical treatment, 4
    treatment for strychnine poisoning, 53
    useful mechanism, 105
Vomiting nuts, 24

W

Wa-Hoo Bitters, 204
Wainewright, Elisa, 113–114, 116, 118
Wainewright, Thomas Griffiths, 112–123
Wakley, Thomas, 87, 168, 212

Walckiers, Vicomte Edouard de, 35
Walker, Caroline, 199
Wallonia, 90
Ward, Colonel, 139, 146
Warder, Dr. Alfred, 164
Warner, Dr. W.C., 17
Warner Lambert Healthcare, 204
Washington State, 265
Waterhouse, William, 185
Waterloo, Battle of, 247
Waterloo, London, 192, 195, 243, 254
WD-40, 78
Weakness, 4–6, 64
Webster, John, 233
Weekly Despatch, The, 240, 245
Weizmann, Chaim, 223
Wells, H.G., 201
Wells's Tablets, 204
Western Dispensary, 262
Westmoreland Gazette, 120
Westwood, Alfred, 43
Wheeldon family, 240–245
Wilkins, Serjeant, 149, 164
Williams, John, 121
Willstätter, Richard, 224
Wilson's Wa-Hoo Bitters, 204
Windham, William, 167
Wireless experiments, 231
Witham, Jane, 145, 178

Wodehouse, P.G., 254–255
Wolves, poisoning of, 25
Women, diseases of, 5
Wonder drug, strychnine as, 267
Woodhouse, Richard, 121
Woodward, Margaret, 234
Woodward, Robert Burns, 227,
    234–237, 264
Wooler, 95
Working classes, diseases of, 6
World War One, *see* First World War
Wormley, T.G., 165
Wright, Edwin, 128
Wrightson, Dr. Francis, 157

Y

Yeomans, William, 111
Yields, synthetic, 236–237
Young, Chief Inspector, 256–257

Z

Zaytoun, 19
Zimmermann, Captain, 35
Zinc, 199